GEOLOGICAL STRAIN ANALYSIS
A Manual for the $R_f/\emptyset$ Method

Cheeney
1985 December

Related Pergamon Titles of Interest

Books

J. A. E. ALLUM
Photogeology and Regional Mapping

J. G. C. ANDERSON
Field Geology in the British Isles (A Guide to Regional Excursions)

J. G. C. ANDERSON
The Structure of Western Europe

J. G. C. ANDERSON & T. R. OWEN
The Structure of the British Isles, 2nd edition

*D. H. MALING
Measurements from Maps (The Principles and Methods of Cartometry)

T. R. OWEN
The Geological Evolution of the British Isles

N. J. PRICE
Fault and Joint Development in Brittle and Semi-brittle Rock

J. L. ROBERTS
Introduction to Geological Maps and Structures

B. SIMPSON
Geological Maps

Journals

Computers & Geosciences
Journal of African Earth Sciences
†Journal of Southeast Asian Earth Sciences
Journal of Structural Geology
Quaternary Science Reviews

Full details of all Pergamon books and journals and a free specimen copy of any Pergamon journal available on request from your nearest Pergamon office.

*In preparation
†New

GEOLOGICAL STRAIN ANALYSIS
A Manual for the $R_f/\emptyset$ Method

by

RICHARD J. LISLE
Department of Geology
University College of Swansea, U.K.

PERGAMON PRESS

OXFORD · NEW YORK · TORONTO · SYDNEY · FRANKFURT

U.K.	Pergamon Press Ltd., Headington Hill Hall, Oxford OX3 0BW, England
U.S.A.	Pergamon Press Inc., Maxwell House, Fairview Park, Elmsford, New York 10523, U.S.A.
CANADA	Pergamon Press Canada Ltd., Suite 104, 150 Consumers Road, Willowdale, Ontario M2J 1P9, Canada
AUSTRALIA	Pergamon Press (Aust.) Pty. Ltd., P.O. Box 544, Potts Point, N.S.W. 2011, Australia
FEDERAL REPUBLIC OF GERMANY	Pergamon Press GmbH, Hammerweg 6, D-6242 Kronberg-Taunus, Federal Republic of Germany

Copyright © 1985 R. J. Lisle

All Rights Reserved. No part of this publication may be reproduced, stored in a retrieval system or transmitted in any form or by any means: electronic, electrostatic, magnetic tape, mechanical, photocopying, recording or otherwise, without permission in writing from the publishers.

First edition 1985

Library of Congress Cataloging in Publication Data

Lisle, Richard J.
Geological strain analysis.
On t.p. "f" is subscript.
Bibliography: p.
Includes indexes.
1. Rock deformation. 2. Strain and stresses.
3. Strain gages. I. Title.
QE604.L57 1985   551.8   85-16952

British Library Cataloguing in Publication Data

Lisle, Richard J.
Geological strain analysis : a manual for
the $R_f/\emptyset$ method.
1. Rocks — Testing   2. Strains and stresses
I. Title
552'.06   QE431.6.M4
ISBN 0-08-032590-4 Hardcover
ISBN 0-08-032589-0 Flexicover

Printed in Great Britain by A. Wheaton & Co. Ltd., Exeter

Contents

	LIST OF SYMBOLS	vii
1.	INTRODUCTION	1
	1.1 Strain analysis	1
	1.2 Uses of strain analysis	1
	1.3 Elliptical objects as strain markers	2
2.	THEORETICAL BASIS OF THE R_f/ϕ METHOD	3
	2.1 Deformation of a passive elliptical marker	3
	2.2 Suites of markers with identical initial eccentricity	4
	2.3 Suites of markers with identical initial orientation	6
3.	DATA COLLECTION	8
	3.1 Suitable material for R_f/ϕ analysis	8
	3.2 Making the R_f/ϕ measurements	8
	3.3 Amount of data required	10
	3.4 Plotting the $R_f\phi$ data	11
	3.5 Three-dimensional strain	11
4.	APPLICATION OF THE STANDARD CURVES TO DETERMINE THE STRAIN RATIO	12
	4.1 Introduction	12
	4.2 The Symmetry Test	12
	4.3 The θ-Distribution Test	15
	4.4 The θ-Distribution Test for low Ri particles	17
	4.5 Destraining procedure assuming a symmetrical theta distribution	20
	4.6 Initial fabrics symmetrical about bedding	20
5.	EXTENSIONS OF THE R_f/ϕ METHOD	23
	5.1 Introduction	23
	5.2 Viscosity contrast between the inclusions and matrix	23
	5.3 Deformation by pressure solution	25
	5.4 Deformation of anisotropic markers	25
	5.5 The R_f/ϕ method applied to heterogeneously strained rocks	25
6.	THE MARKER DEFORMATION GRIDS	26
7.	PURE SHEAR DEFORMATION PATHS FOR VARIOUS VISCOSITY CONTRASTS	78
	APPENDIXES	
	1. Basic equations	39
	2. Production of symmetrical R_f/ϕ patterns from asymmetrical initial fabrics	
	REFERENCES	91
	AUTHOR INDEX	97
	SUBJECT INDEX	99

List of Symbols

R	the axial ratio of an ellipse, i.e. long axis length/short axis length ($R \geq 1.0$)
ε	$\tfrac{1}{2} \ln R$
R_i	the initial or undeformed axial ratio of an elliptical strain marker
R_f	the final or deformed axial ratio of an elliptical strain marker
R_s	the strain ratio, i.e. the ratio of the principal stretches for the strain ellipse
$\varepsilon_i, \varepsilon_f, \varepsilon_s$	$\tfrac{1}{2} \ln (R_i, R_f, R_s)$
θ	the angle between the long axis of a marker and the maximum extension direction in the intitial or undeformed state
$\varnothing$	the angle between the long axis of a marker and the maximum extension direction in the final or deformed state
$\theta', \varnothing'$	the equivalents of $\theta, \varnothing$ measured with respect to a reference line which does not correspond to the maximum extension direction
V	the ratio of the viscosity of inclusions to that of the whole rock system, $\mu_{\text{inclusions}} / \mu_{\text{whole rock}}$.
H	the harmonic mean of the R_f values of a suite of elliptical markers
$\bar{\varnothing}, \bar{\varnothing}'$	the vector mean of the $\varnothing, \varnothing'$ angles respectively

1
Introduction

1.1 STRAIN ANALYSIS

Structural geology is concerned with deformation phenomena in rocks and the branch of structural geology which deals with the quantification of geological deformation is known as geological strain analysis. The strain analyst devises and applies methods to calculate the distortions a volume of rock has been subjected to since its formation.

In most cases the rocks to be analysed have been strained during periods of deformation in the distant past so that we are only able to observe the rocks in their distorted condition. The fact that we do not see the rocks in their undeformed state hampers the analysis of strain as the latter relies on a comparison being made between geometrical form of objects as they were before deformation with their present configuration. In fact our ignorance of the undeformed aspect of a rock would render the task of calculating rock distortion impossible were it not for certain objects within deformed rocks whose shape or geometrical arrangement prior to straining is known or can be assumed. These objects, known as strain markers provide the raw data for strain analysis and include fossils, grains in metamorphic rocks, clasts in sedimentary rocks and sedimentary structures. They range in scale from the distorted arrangement of volcanic centres used to deduce continent-scale deformations (Windley and Davies 1978) to the disrupted form of rutile needles used to quantify the strain in the enclosing quartz grain (Mitra 1978). These strain markers come in a huge variety of forms but have in common that they record either changes of angle between lines or changes in line length resulting from geological strain.

The wide diversity of natural strain gauges which exists in tectonites has meant that a large variety of methods have had to be devised to analyse them. Although attempts to estimate strain date from as far back as the middle of last century it is only in the last twenty years that we have seen the establishment of strain analysis as an important discipline within the field of structural geology. The upsurge of interest and proliferation of techniques coincided with the appearance of J.G. Ramsay's text on quantitative structural geology, 'Folding and Fracturing of Rocks' which provided a clear description of the basic principles upon which strain analysis methods are based.

1.2 THE USES OF STRAIN ANALYSIS

A knowledge of the state of strain in a tectonite is vital if any form of reconstruction of the pre-tectonic geometry of the rock body is to be attempted. Strain analysis provides the basic data required for the restoration of stratigraphic thicknesses (Ramsay 1969), for returning sedimentary basins and structures to their original form or for taking off the distortions produced by folding (e.g. Oertel 1974, Schwerdtner 1977, Cobbold 1980). Ragan and Sheridan (1972) and Sparks and Wright (1979) have used strain analysis to restore sequences of volcanic rocks to their pre-compaction thicknesses. Estimates of the magnitude of strain are important in the study of the rheological properties of rock materials at the time of their deformation. By examining the way strain intensity depends on microstructural parameters such as grain size, useful deductions can be made concerning the deformation mechanisms which operated. Examples of these deductions are provided by the work of Evans and others (1980), Etheridge and Vernon (1981) and Lisle and Savage (1983). Finally, strain data can help throw light on the mode of development of a multitude of geological structures. For instance, discussions on the origin of secondary foliations and crystallographic fabrics have revolved around their supposed relationship to the finite strain in the rock.

1.3 ELLIPTICAL OBJECTS AS STRAIN MARKERS

Strain analysis involves converting the information on length changes and angular distortions provided by strain markers to a more readily understood representation of the state of strain, the strain ellipsoid. The strain ellipsoid, the shape adopted after deformation by an imaginary mechanically passive sphere embedded in the rocks, allows the intensity of the strain and its three dimensional character to be directly appreciated. Clearly ideal strain markers would be objects with an initial spherical shape. Their final shape and orientation would be a direct portrayal of the strain ellipsoid and their "analysis" would be trivial. Although this simple form of analysis has been used for some markers (such as ooids), the rarity of spherical objects has meant that it is of limited potential. On the other hand, a large variety of markers have shapes which, in their undeformed state, approximate to ellipsoids and methods designed to process data from this type of marker have much wider application.

Since these markers possess an eccentricity before straining, their deformed shapes do not directly reflect the shape of the strain ellipsoid. Their final axial ratios and orientations are due to the combined effects of their pre-tectonic eccentricity (often referred to as the "shape factor") and the superimposed strain.

Ramsay (1967, p. 202-211) devised a method of strain analysis using such markers. He showed theoretically that a set of elliptical markers with identical initial eccentricity but variable orientation will show a characteristic pattern if their deformed axial ratios (R_f) and orientations ($\emptyset$) are plotted graphically. The pattern shown by the markers on the $R_f/\emptyset$ plot was shown to be a function of the strain ellipse shape and of the initial eccentricity of the markers. Ramsay's $R_f/\emptyset$ method allowed the effects of initial shape to be distinguished from those due to tectonic strain.

Dunnet (1969) gave a comprehensive description of this method and details of the practical procedures to be followed when applying it to a variety of rock-types.

The $R_f/\emptyset$ technique has since become the most widely employed method with over one hundred applications of the technique described in the literature. In comparative studies of various strain techniques (Hanna and Fry 1979, Siddans 1980, Percevault and Cobbold 1982, Paterson 1983) the $R_f/\emptyset$ technique emerges as one of the most reliable ways of estimating the tectonic strain.

The technique explained in this manual is essentially the Ramsay-Dunnet method but is modified to incorporate the improvements suggested by Lisle (1977b).

2
Theoretical Basis of the $R_f/\emptyset$ Method

2.1 DEFORMATION OF A PASSIVE ELLIPTICAL MARKER

We consider the homogeneous deformation of an elliptical object which is straining as its matrix so that its behaviour can be said to be passive (Fig. 2.1). After a strain of magnitude R_s the elliptical shape (R_i, θ) transforms to one with orientation $\emptyset$ and shape R_f given by

$$\tan 2\emptyset = \frac{2 R_s (R_i^2 - 1) \sin 2\theta}{(R_i^2 + 1)(R_s^2 - 1) + (R_i^2 - 1)(R_s^2 + 1) \cos 2\theta} \tag{2.1}$$

$$R_f = \left[\frac{\tan^2\emptyset (1 + R_i^2 \tan^2\theta) - R_s^2 (\tan^2\theta + R_i^2)}{R_s^2 \tan^2\emptyset (\tan^2\theta + R_i^2) - (1 + R_i^2 \tan^2\theta)}\right]^{\frac{1}{2}} \tag{2.2}$$

These equations are derived by Ramsay (1967, p. 205-209) and other equations relating the five variables (R_i θ R_s R_f and $\emptyset$) are listed in Appendix 1.

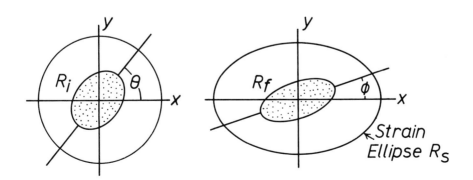

Fig. 2.1. Straining of an elliptical marker

As a result of strain the long axis of the ellipse rotates towards the maximum extension direction (X), i.e. $\theta > \phi$. The same is true for a passive line whose rotation obeys the Harker Equation:

$$\tan \phi_{line} = \tan \theta_{line} \cdot 1/R_s \tag{2.3}$$

$$\text{or } \tan 2\phi_{line} = \frac{2 \tan \theta_{line}}{R_s - \frac{1}{R_s} \tan^2 \theta_{line}} \tag{2.4}$$

Re-arranging eq. 2.2 to facilitate a comparison with eq. 2.4 leads to

$$\tan 2\phi = \frac{2 \tan \theta}{\frac{R_i^2}{R_i^2-1} \left[(R_s - \frac{1}{R_s} \tan^2 \theta) - \frac{1}{R_i^2} \left(\frac{1}{R_s} - R_s \tan^2 \theta \right) \right]} \tag{2.5}$$

Since eq. 2.4 is the limit of equation 2.5 as R_i approaches infinity, we can conclude that a passive line rotates in response to an imposed strain in a similar fashion to an ellipse of very large initial axial ratio. This becomes important in Chapter 4, when we consider the relative rotation of bedding and elliptical markers on a planar section.

2.2 SUITES OF MARKERS WITH IDENTICAL INITIAL ECCENTRICITY

A group of ellipses of identical initial shape deforms to give ellipses of variable R_f and ϕ owing to their dissimilar initial orientations (Fig. 2.2).

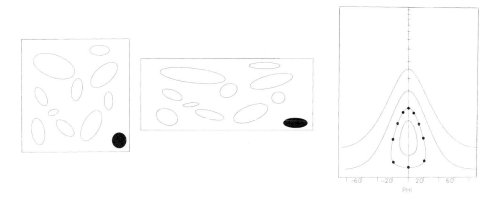

Fig. 2.2. Markers with constant R_i

If deformed markers belonging to such a suite are plotted on a graph of R_f against ϕ they define a curve given by

$$\cos 2\phi = \frac{(R_f + 1/R_f)(R_s + 1/R_s) - 2(R_i + 1/R_i)}{(R_f - 1/R_f)(R_s - 1/R_s)} \tag{2.6}$$

with R_s and R_i as constants. Several such curves are shown in Fig. 2.2. They demonstrate that a constant R_i suite of ellipses will show a definite range of final shapes. The aspect ratios of the extreme shapes (Rf_{max}, Rf_{min}) are simple products or quotients of R_i and R_s since they result from the parallel or perpendicular superimposition of these shape components.

Theoretical Basis of the $R_f/\emptyset$ Method

$$R_{f_{max}} = R_s R_i$$

$$R_{f_{min}} = \text{the greater of } \frac{R_s}{R_i}, \frac{R_i}{R_s}$$

Fig. 2.3 shows how the magnitude of the extreme marker shapes depends on R_i and R_s.

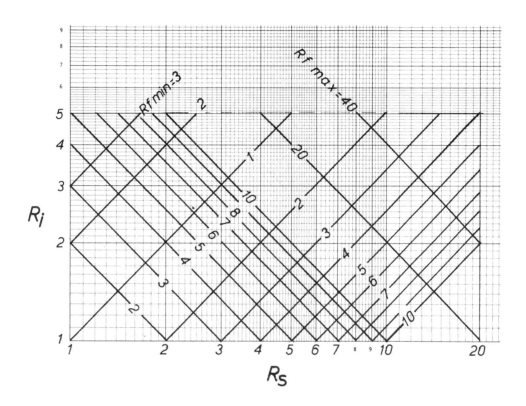

Fig. 2.3. The maximum and minimum axial ratios from a suite of deformed markers

Some R_i curves in Fig. 2.2 span a restricted range of $\emptyset$ values indicating that the deformed markers possess a limited spread of orientations. The maximum angular interval over which orientations occur is known as fluctuation (Cloos 1947, Ramsay 1967, p. 207). The size of this angle ($2\emptyset_{max}$) is given by the equation

$$\sin 2\emptyset_{max} = \frac{R_i - 1/R_i}{R_s - 1/R_s} \tag{2.7}$$

which is equivalent to Dunnet's (1969) eq. 32 expressed in terms of axial ratios. For $R_i > R_s$, $2\emptyset_{max}$ is $180°$, i.e. the fluctuation is unrestricted. By differentiating equation 2.6 and equating to zero we obtain the R_f value of the marker which has the orientation $\emptyset_{max}$:

$$R_{f(\emptyset_{max})} = \frac{R_s + 1/R_s}{R_i + 1/R_i} + \left[\left(\frac{R_s + 1/R_s}{R_i + 1/R_i}\right)^2 - 1\right]^{\frac{1}{2}} \tag{2.8}$$

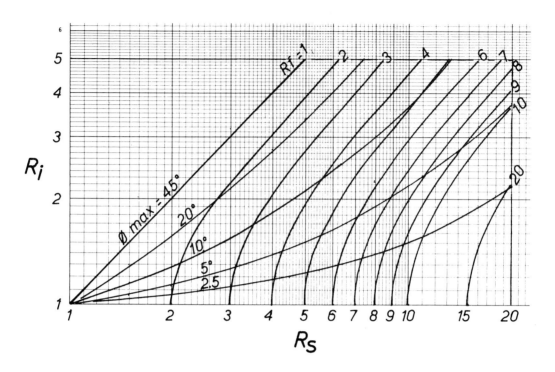

Fig. 2.4. R_f and $\emptyset$ for markers with extreme $\emptyset$ angles

Fig. 2.4 shows how the shape and orientation of the markers with extreme orientations are determined by R_i and R_s.

2.3 SUITES OF MARKERS WITH IDENTICAL INITIAL ORIENTATION

Markers sharing the same initial orientation (Fig. 2.5) define a curve on an $R_f/\emptyset$ diagram termed a theta-curve (Lisle 1977b). By varying the initial orientation of the whole suite a family of theta-curves are produced which radiate from the point

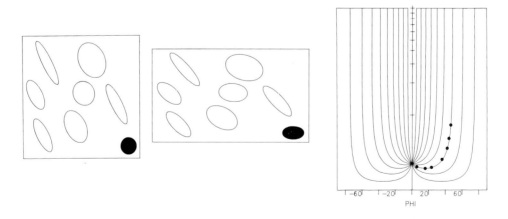

Fig. 2.5. Markers with constant θ

($\emptyset = 0$, $R_f = R_s$). They are drawn by substituting the appropriate values of R_s and θ into the equation (Lisle 1977b)

Theoretical Basis of the R_f/ϕ Method

$$R_f = \left[\frac{\tan 2\theta \, (R_s^2 - \tan^2\phi) - 2R_s \tan\phi}{\tan 2\theta \, (1 - R_s^2 \tan^2\phi) - 2R_s \tan\phi} \right]^{\frac{1}{2}} \tag{2.9}$$

To draw the $\theta = 45°$ curve, use is made of

$$R_f = \left[\frac{\tan^2\phi - R_s^2}{R_s^2 \tan^2\phi - 1} \right]^{\frac{1}{2}} \qquad \text{Dunnet \& Siddans (1971)}$$

When θ is greater than $45°$, the curves have a minimum at a ϕ value obtained by differentiating equation 2.9 and equating $\frac{dR_f}{d\phi}$ to zero. This yields

$$\tan \phi_{(\min R_f)} = \frac{1}{2R_s} \left[\tan 2\theta (R_s^2 + 1) \pm (\tan^2 2\theta (R_s^2 + 1)^2 + 4R_s^2)^{\frac{1}{2}} \right] \tag{2.10}$$

When the family of R_i curves and the family of θ-curves are plotted together, they form a net on the R_f/ϕ diagram. These "marker deformation nets" are given in Chapter 6. We can see that they vary in shape according to the strain ratio and form a rectangular grid when $R_s = 1$. They allow us to read off the deformed characteristics (R_f, ϕ) of an object with initial parameters R_i, θ and vice versa. Consequently, if we place the grid on an R_f/ϕ data set, we can read off the axial ratios and orientations we would obtain by 'de-straining' the rock by an amount corresponding to the R_s value of the grid.

The essential feature of the R_f/ϕ method is that we, (by using the standard nets in Chapter 6) graphically de-strain the markers repeatedly until the group of restored markers shows R_i and θ distributions which most closely matches those we assume existed in the rock in the undeformed state.

3
Data Collection

3.1 SUITABLE MATERIAL FOR $R_f/\emptyset$ ANALYSIS

A major advantage the $R_f/\emptyset$ method possesses is its ability to estimate strain in a wide range of rock types. The basic requirement a rock must meet for the application of this method is the presence of elliptical, subelliptical or parallelogrammatical inclusions the boundaries of which have remained attached to the same material points during the deformation. Inclusions whose boundaries have migrated through the rock during or after deformation are not suitable strain markers. Examples of the latter are grains which achieve a new configuration resulting from grain growth or cataclastic breakdown.

In spite of this limitation the method has proven itself versatile and a wide range of rock types are amenable to this form of analysis. Table 3.1 is a compilation of analyses performed on various lithologies. The listed published analyses contain useful supplementary details of the method as applied to particular rock-types.

3.2 MAKING THE $R_f/\emptyset$ MEASUREMENTS

Measurements should be made on planar surfaces; a joint face at the outcrop, a sawn surface on a hand specimen or a thin section. The attitude of these planes relative to geographical reference directions should be carefully recorded. If it is intended to determine the three-dimensional strain, three planes are in general required (see Section 3.5). If the orientation of these planes can be selected, as in the case of cut faces on a hand specimen, then it is preferable if they are orthogonal or approximately so (see Owens 1984).

Measurement of axial ratios and orientation can be made directly on the rock surface, from photographs or from traced drawings of the marker outlines. Paterson (1983) has compared these various options and found that the edges of individual particles can be difficult to discern on photographic images. Elliott (1970) and Milton and Chapman (1979) made use of peels from etched surfaces on conglomerates. In thin sections Mukhopadhyay (1973) and Tan (1976) made R_f and $\emptyset$ measurements using the microscope's mechanical stage and an ocular micrometer. Dunnet (1969) placed thin sections directly in the enlarger to make large photographs for measurements. Alternatively, large drawings can be made of the markers' outlines by projecting the thin section on to paper.

To measure R_f, the dimensions of the long and short axes are measured and the ratio formed. For shapes which deviate from perfect ellipses, Holst (1982) took the

TABLE 3.1

Applications of the R_f/ϕ technique

Conglomerates	Brun and others (1981), Chaudnuri and Pal (1980), Deramond and Rambach (1979), Dunnet (1969, Gay (1969), Hutton (1979), Kelly and Max (1979), Le Corre and Le Theoff (1976), Mukhopadyay and Bhattacharya (1969), Roy and Faerseth (1981), Stauffer and Burnett (1979), Stephens (1975).
Breccias	Kligfield and others (1981), Tobisch and others (1979)
Oolites	Boulter (1976), Dunnet (1969), Hanna and Fry (1979), Pfiffner (1980), Ramsay and Huber (1983), Tan (1976).
Oncalites	Kligfield and others (1981).
Quartz grains	Law and others (1984), Moncktelow (1981), Marjoribanks (1970), Mukhopadyay (1973).
Quartz aggregates in gneisses	Jensen (1984), Odling (1984).
Biotite/muscovite aggregates	Gray and Durney (1979).
Cordierite crystals and pseudomorphs	Brun and others (1981), Robin (1977).
Feldspar clasts	Borradaile (1979).
Pyrites	Sen and Mukherjee (1972).
Amygdales	Ribeiro and others (1983).
Varioles in basic volcanics	Barr and Coward (1974).
Xenoliths	Coward (1976).
Pumice- and accretionary lapilli	Bell (1981), Boulter (1983), Roberts and Siddans (1971), Sparks and Wright (1979).
Boudins	Sen and Mukherjee (1972).
Burrows and pipes	Geiser (1974), Wheeler (1984).
Nodules in slates	Martin-Escorza and Martin-Montalvo (1980).
Reduction spots	Graham (1978), Siddans (1979).
Polygonal desiccation cracks	Harvey and Ferguson (1981).

longest dimension as the major axis measurement and the longest dimension perpendicular to the major axis as the shortest axis. Rectangular, rhombic and pear-shaped particles were measured by Dunnet (1969) by estimation of the axes of the ellipse of equivalent area.

Roder (1977) has drawn attention to the fact that a unique ellipse can be inscribed within a parallelogram and thus that a wide range of shapes including polygons can be used as strain gauges for the $R_f/\emptyset$ method. Figure 3.1 shows a variety of objects which, even when of unknown initial shape, contain equivalent information to a deformed ellipse and can therefore function as strain markers in this method.

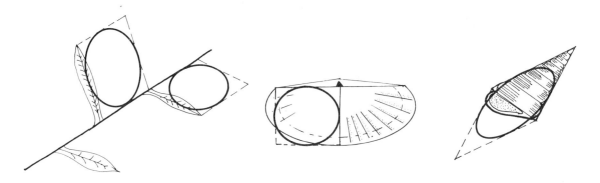

Fig. 3.1. Various-shaped objects which can function as elliptical strain markers

Harvey and Ferguson (1981) show how a single unique ellipse can be assigned to any convex polygon and apply this to the analysis of strained desiccation crack polygons.

The directions of long axes of the elliptical shapes, being non polar, can be expressed as an angle in the range $+90°$ to $-90°$ with respect to a reference line on the plane. These angles are $\emptyset'$ angles. It is convenient to choose a reference line in the general direction of preferred orientation of the markers. The paired values of $\emptyset'$ and the associated R_f value should be tabulated to facilitate plotting.

In automated measurement procedures, R_f and $\emptyset$ can be calculated directly from the co-ordinates of a number of points on the marker's outline. Paterson (1984) uses four such points and presents the relevant equations for R_f and $\emptyset$. Siddans (1976) digitises more points per marker and from this data the computer is able to select only those outlines which are sufficiently elliptical in shape. The other, non-elliptical shapes are discarded from the analysis. Little is known however about errors incurred by the use of markers which deviate in shape from exact ellipses, so setting limits on the amount of acceptable non-ellipticity remains somewhat arbitrary.

Where inclusions of different lithologies are present care should be taken to separate data and analyse each fraction separately. Avoid sampling markers from heterogeneous deformed specimens where the direction of preferred orientation of the long axes varies visibly across the sampled domain. A number of studies (e.g. Lisle 1977b, Lisle and Savage 1983) have shown that marker shapes are related to their size. This indicates a non-passive strain response by the markers and can be detected effectively by plotting a graph of long axis and short axis dimensions (see Ramsay 1967, p. 193, Elliott 1970). Figure 3.2 illustrates both possible types of size dependence related to different operating mechanisms of deformation on the scale of the markers. A positive correlation between size and shape can be produced by crystal-plastic deformation of the markers (Fig. 3.2a) and a negative correlation has been observed in rocks where the markers have been subject to pressure solution deformation (Fig. 3.2b). Where such size effects are evident, the different size fractions should be analysed separately.

3.3 AMOUNT OF DATA REQUIRED

Although the method allows an estimate of the strain to be made on the basis of as little as 10 measurements, larger sample sizes allow checks to be applied to the assumptions involved in the method. Dunnet (1969) recommended minimum numbers of markers depending on the degree of eccentricity shown by the objects in the stacking material. For example, for ooids he recommends a minimum of 30-40 as against 60-100 for the originally more eccentric markers such as pebbles. The variant of the method described here works equally well on shapes of low and high initial eccentricity. The way the present technique is devised allows a wide range of sample sizes to be

Data Collection

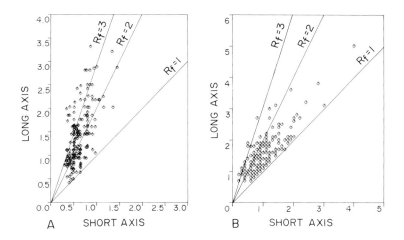

Fig. 3.2. Shape (R_f) related to size of strain markers
A. Plastically deformed quartz grains Angers, France (data: Bouchez 1977)
B. Quartz grains deformed by pressure solution Aberystwyth Grits, Wales

handled. For large samples, more sensitive tests can be applied. The minimum number of data required for the various tests are 10, 25, 50 and 100 respectively. Borradaile (1984) has investigated the question of optimum sample sizes for this $R_f/\emptyset$ technique and finds that sensitivity improves markedly as the sample size increases to around 50 but after 75 further improvement becomes much less pronounced.

3.4 PLOTTING THE $R_f/\emptyset$ DATA

Markers are plotted as points on a graph of R_f against $\emptyset'$. To allow a comparison between the point distribution obtained by plotting $R_f/\emptyset$ data pairs and the standard curves given in Chapter 5, it is important to scale the axes as follows: 10 degrees = 1 cm on $\emptyset'$ axis, R_f axis logarithmic with a 12.5 cm cycle. Graph paper with these scales is widely available (e.g. Chartwell D.5521). The data should be plotted on a transparent overlay or on transparent graph paper.

3.5 THREE DIMENSIONAL STRAIN

The orientation and axial ratios of the strain ellipsoid are calculated from the strain ratios, determined on a number of planes through the rock. For the calculations involved in the combination of two-dimensional strain data to build up the ellipsoid the reader is referred to Ramsay (1967, p. 142-149), Shimamote and Ikeda (1976), Oertel (1978), Miller and Oertel (1979), Milton (1980), Siddans (1980) and Owens (1984).

Calculations are very much simplified if measurements are made on planes which are principal planes of the strain ellipsoid, though deciding on the attitude of the principal planes may not always be easy.

4
Application of the Standard Curves to Determine the Strain Ratio

4.1 INTRODUCTION

After the data has been plotted on an $R_f/\emptyset$ diagram the analysis proceeds by comparing the data point distribution with the standard curves reproduced in Chapter 6. Strain magnitude is determined by finding the set of curves which according to some specific criterion most closely fits the data points. There are a number of criteria which can be applied to the fitting procedure, each based on its own set of assumptions about the deformation and the starting material. Sometimes even the best-fitting curves are a rather poor fit to the data and this suggests that the assumptions implied by the curve-fitting criteria are suspect. This feature, the ability to test its own assumptions, represents an important advantage that the $R_f/\emptyset$ method has over other strain methods. If one set of assumptions is found unsatisfactory another fitting procedure based on different assumptions can be used.

The flow chart (Fig. 4.1) sets out the sequence to be followed when the strain ratio is to be determined from the standard curves. The various routes through the flow chart correspond to the different forms of analysis that are applied to rocks with viscosity contrasts, to those with primary fabrics etc. The tests that are applied to the data are described in the following sections.

4.2 THE SYMMETRY TEST

A group of markers with random orientations before straining will tend to show a symmetrical $R_f/\emptyset$ pattern after straining (Fig. 4.2a). An initial preferred orientation of marker long axes will usually, but not always,* produce a pattern which is asymmetrical about the mean $\emptyset'$ line on the diagram. Figure 4.2b shows $R_f\emptyset$ data derived from a unimodal distribution of initial orientations. The data points are strung out along a theta curve (see section 2.3) and generally show therefore an asymmetry.

---

*Symmetrical $R_f/\emptyset$ patterns can result from symmetrical initial fabrics where the strain is imposed coaxially and, fortuitously, by the oblique straining of an asymmetrical initial fabric (see Appendix 2).

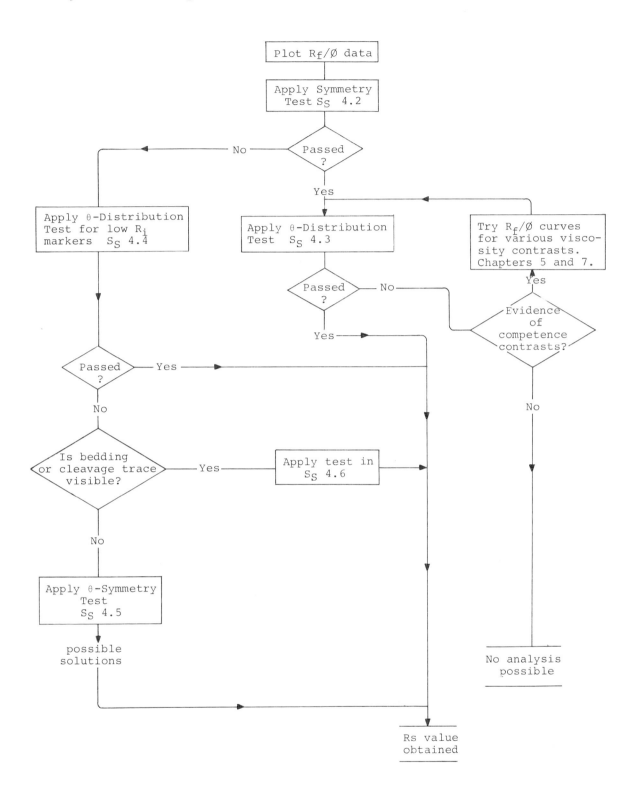

Fig. 4.1 Flow chart for the procedures leading to the determination of the strain

Application of the standard curves to determine the strain ratio

If therefore we can assess the degree of symmetry, we are able to make certain decisions regarding the undeformed rock. Different ways of assessing asymmetry have been proposed (e.g. Ramsay and Huber 1983, p. 82-83) and the one explained below is similar to that proposed by Dunnet and Siddans (1971). A pocket calculator will be useful to calculate the vector mean $\emptyset'$ and the harmonic mean of the R_f values. These are calculated from the following equations

$$\text{Vector mean } \emptyset \text{ or } \bar{\emptyset} = \tfrac{1}{2}\arctan(\Sigma \sin 2\emptyset / \Sigma \cos 2\emptyset)$$

$$\text{Harmonic mean } H = N / (R_{f_1}^{-1} + R_{f_2}^{-1} + R_{f_3}^{-1} \ldots + R_{f_N}^{-1})$$

These means correspond to the vertical and horizontal lines drawn on the diagram in Fig. 4.2b which divide the diagram into four areas labelled A, B, C and D. An index of symmetry can be defined as

$$I_{SYM} = 1 - (|n_A - n_B| + |n_C - n_D|)/N$$

where n_A, n_B, n_C, n_D are the number of points occurring in areas A, B, C, D respectively and N is the total number of data points plotted.

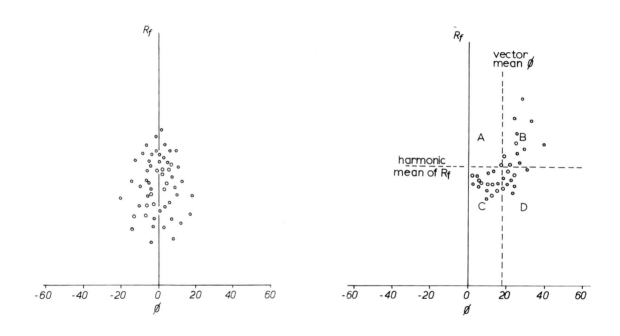

Fig. 4.2

A. Initial random fabric

B. Strong initial preferred orientation of marker long axes

High values of I_{SYM} indicate a highly symmetrical pattern of data points. Low values of I_{SYM} will make the hypothesis of no initial fabric suspect. Two hundred trial calculations of I_{SYM} for randomly sampled markers from a uniform orientation distribution were made for various sample sizes (N) and strains (R_s). Only 5% (and 10% in brackets) of the random samples gave lower values of I_{SYM} than those in Table 4.1. These values can be used therefore as critical values for the test. If the I_{SYM} is less than the appropriate critical value in Table 4.1 we can conclude that the markers did not come from a uniform orientation distribution (with the knowledge that in making this decision, we have a 1 in 20 (10) chance of being wrong). It is important to note from the values in Table 4.1 that asymmetrical patterns can occur even when an initial rock fabric was absent and that this is particularly marked for small

Geological Strain Analysis   15

$R_f/\emptyset$ samples. Dunnet and Siddans (1971) and Bell (1981) have used a critical value of 0.9 for a similar index of symmetry.

On the basis of this symmetry check it is not possible to distinguish between a rock without an initial fabric and one in which the strain has been symmetrically superimposed on an initial fabric. A symmetrical $R_f/\emptyset$ plot could be interpreted in either way.

		Sample Size N				
		20	35	60	100	200
	1.5	0.3	0.51	0.60	0.74	0.82
		(0.4)	(0.63)	(0.67)	(0.78)	(0.85)
	2.0	0.5	0.63	0.73	0.80	0.86
		(0.5)	(0.63)	(0.77)	(0.82)	(0.88)
Rs	3.0	0.5	0.63	0.73	0.80	0.87
		(0.6)	(0.63)	(0.77)	(0.82)	(0.88)
	5.0	0.5	0.63	0.73	0.82	0.87
		(0.6)	(0.63)	(0.77)	(0.82)	(0.88)
	10.0	0.6	0.63	0.73	0.82	0.87
		(0.6)	(0.63)	(0.77)	(0.84)	(0.89)

Table 4.1 Critical values of I_{SYM} used in the Symmetry Test. The values shown are the 5% (10%) percentage points of the I_{SYM} distribution.

4.3 THE θ-DISTRIBUTION TEST

For data judged by the test in section 4.2 to be symmetrical we proceed by comparing the data points to the shape of the theta curves on the standard charts.

If we take any set of theta curves from Chapter 6 (say $R_S = 1.5$, Fig. 4.3b) and superimposed our data points on it, we can read off a theta value for each marker. This is the orientation that marker would adopt if we were to de-strain the rock by a strain of $R_S = 1.5$. By considering all the data points and their spread with respect to the theta curves we obtain an impression of the distribution of long axes in the de-strained state (Fig. 4.3b). The principle behind the θ-distribution test is to find the set of theta curves for which the distribution of θ's is most uniform (Fig. 4.3c). The R_S value associated with this best-fit set of curves is the best estimate we can make of the strain ($R_S = 3.0$ in this example).

To avoid working through the whole selection of standard graphs it is efficient to have a rough indication of the strain ratio so that only a few of the curves need be consulted. To provide an approximate value of R_S, use can be made of the harmonic mean of the axial ratios calculated for the purpose of the symmetry test (section 4.2). It has been shown (Lisle 1977a, Lisle 1979) that the harmonic mean (H) approximates, and slightly overestimates the strain ratio (Fig. 4.4 shows how H varies as a function R_S and R_i). The curves are based on a model involving passive deformation of variably-oriented ellipses. This chart can be used to estimate R_S, especially if the likely average R_i value can be judged from the nature of the rock and the inclusions. Table 4.2 gives average R_i values determined in published $R_f/\emptyset$ analyses on various rock types and could assist in assigning a likely average R_i.

Application of the standard curves to determine the strain ratio

Other fast estimates of the strain ratio use the extreme points on the R_f/ϕ diagram. Either $R_{f_{max}}$ and $R_{f_{min}}$ can be used (Fig. 2.3) or the orientation and axial ratio of the marker with extreme orientation, ϕ max (Fig. 2.4).

Although a visual comparison of data with the standard curves is probably adequate in most cases, it sometimes arises that even the best-fitting theta curves do not succeed in dividing the data points very evenly. The application of a statistical test, the χ^2 test, allows the goodness of fit of the curves to the data to be quantified and thus allows an objective judgement to be made as to the relevance of the assumed model of deformation.

χ^2 is calculated as follows:

$$\chi^2 = \Sigma(O-E)^2/E$$ where O is the observed number of points occupying a cell bounded by two theta curves and E is the number expected in that cell.

Rock-type/markers	$\bar{R}_i$	Source
Sandstone/quartz grains	1.45-1.55	Griffiths 1967 p. 124
	1.35-1.65	Bokman 1957
Greywacke/slate particles	1.5 -2.0	Dunnet 1969
Conglomerate/pebbles	1.85	Lisle 1977a
	1.5 -2.0	Stauffer & Burnett 197
Grit	1.6 -2.6	Dunnet 1969
/Ooids	1.4	Pfiffner 1980
	1.1 -1.3	Dunnet 1969
	1.2 -1.4	Martin Escorza 1969
Pisolitic tuff	1.38	Dunnet & Siddans 1971
/accretionary lapilli	<1.5	Helm & Siddans 1971
Tuff/pumice clasts	1.7 -2.0	Sparks & Wright 1979
"Pipe" burrows (in bedding-parallel sections)	1.2 -1.4	Wheeler 1984
Strained cordierite crystals	1.4 -1.75	Robin 1977

TABLE 4.2 Initial axial ratios of various marker types

Since the null hypothesis is a uniform distribution of thetas, $E = N/k$, where k is the number of cells used. The test requires that $E \geqslant 5$, so when N is low, k should be reduced by combining cells. Figure 4.5 shows the suggested cell configurations for various values of N.

Critical values of χ^2 are given in Table 4.3.

Low values of χ^2 indicate that the theta curves fit the data well. A value of χ^2 lower than the appropriate critical value in Table 4.3 implies that the R_f/ϕ point distribution is compatible with the model involving passive deformation of random markers. It should be realised that a low χ^2 value does not prove the model's validity. Consider a deformed rock with an initial fabric with R_i/θ characteristics similar to those you would obtain by straining a random fabric. The procedure outlined above will not be able to detect the presence of the initial fabric in the rock. Instead, a fictitious "strain" will be determined which is the product of the superimposition of the true strain on the initial fabric "strain". Although many fabric studies of undeformed sediments have been made, little is known of their R_i/θ properties. More work is required in this direction to supplement that by Boulter (1976), Seymour and Boulter (1979) and Holst (1982).

A value of χ^2 at best-fit which is higher than the critical value indicates that the sample was very probably deformed in a manner other than the way the passive marker model prescribes or that a pre-strain fabric was present. If the latter is the case, the initial fabric must have been of the rather special type (see Appendix 2).

For examples of the type of conclusions to be drawn from the θ-Distribution Test, refer to Lisle (1977b) and Robin (1977).

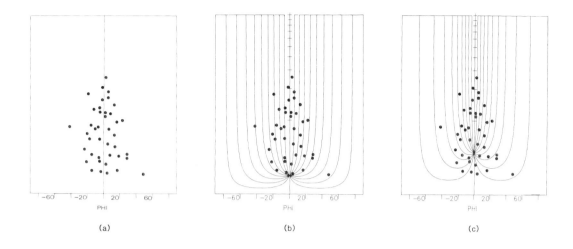

Fig. 4.3 Fitting the theta curves to R_f/ϕ data
(a) The R_f/ϕ data. (b) A poor fit. (c) A good fit.

	10% level	5% level
N = 25-50 (k = 5)	4.605	5.991
N = 51-100 (k = 10)	12.02	14.07
N > 100 (k = 20)	24.77	27.59

TABLE 4.3 Critical values of χ^2 used in the σ-Distribution Test

4.4 THE θ-DISTRIBUTION TEST FOR LOW R_i PARTICLES

Asymmetrical R_f/ϕ clusters suggest the presence of a pre-tectonic fabric in the rock. In clastic sediments however an initial fabric is likely to be less marked for grains of low aspect ratio (Griffiths 1967). If it is reasonable to assume that the low shape fraction of the particles had a random fabric initially, this opens the possibility of applying the θ-Distribution Test to determine the strain ratio. In applying the test we need to be able to exclude the markers with a high R_i from consideration and to fit the θ-curves on the basis of the θ-distribution of low R_i markers, say, those where $R_i < 1.5$. Figure 4.6 shows the type of R_f/ϕ diagram expected if only the high aspect ratio markers show a preferred orientation. The high R_i markers will be strung out along a theta curve and form a "tail" to the main cluster of points. If this shape can be recognised, the data which define the "tail" can be left out of the theta curve fitting procedure.

Clearly the application of this method will require a large quantity of data so that the likelihood of obtaining spurious tails by chance is reduced.

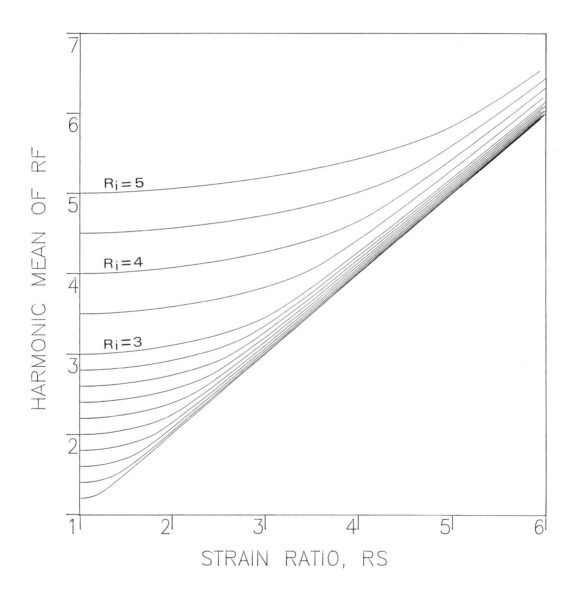

Fig. 4.4 A graph for estimating R_s from the harmonic mean of the axial ratios of deformed passive elliptical markers

Geological Strain Analysis

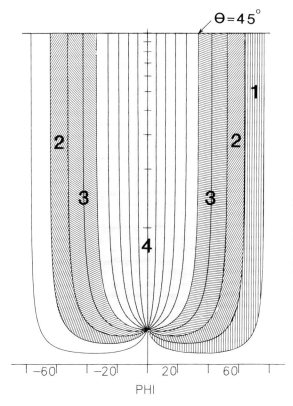

For N > 100, use type 1 cells, k = 20

100 > N > 50, type 2 cells (symmetrical pairs), k = 10

50 > N > 25, type 3 cells, k = 5

For N < 25 use type 4 cells bounded by the ∅ = 45° curve, k = 2

Fig. 4.5 Cell configurations for the χ^2 test

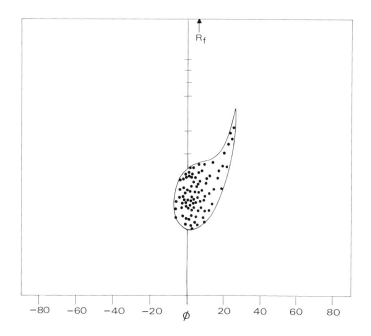

Fig. 4.6 $R_f/\emptyset$ pattern with "tail". High aspect ratio particles showed the strongest preferred orientation before straining

4.5 DESTRAINING PROCEDURE ASSUMING A SYMMETRICAL THETA DISTRIBUTION

If we suspect the existence of a preferred orientation of the markers prior to deformation we will need to make assumptions regarding the nature of this preferred orientation in order to determine the strain. As assumption commonly used is that the pre-tectonic distribution of long axis orientations was symmetrical (Fig. 4.7).

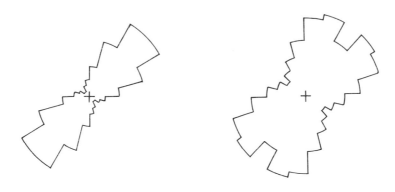

Fig. 4.7 Examples of symmetrical fabrics represented by circular histograms of theta values

The procedure for finding R_s is to compare the $R_f/\emptyset$ points with the marker deformation nets until a symmetrical distribution of thetas is brought about. As we can no longer assume that $\bar{\emptyset}'$ is parallel to the X direction of the strain ellipse, we translate the data parallel to the $\emptyset$ axis during the search for the curves which best fit the data (Fig. 4.8).

If a symmetrical arrangement of θ's can be achieved, R_s, $\emptyset'_X$ at best fit give a possible solution for the strain (its magnitude and orientation respectively). However this may not represent the unique solution and it may require supplementary information, such as knowledge of the orientation of the X axis of the strain ellipse, before the strain can be uniquely determined. Modified versions of this procedure which embody more assumptions or require more data are described below.

4.6 INITIAL FABRICS SYMMETRICAL ABOUT BEDDING

The fabrics of many undeformed sediments show a pattern of preferred orientation which is symmetrical about bedding (Holst 1982). If we choose to make this assumption and can observe the bedding trace on the deformed rock surface, we can select the marker deformation net which yields a distribution of thetas which is symmetrical about bedding. Since bedding can be treated as the point $R_f = 50$, $\emptyset' = \emptyset'_{bedding}$ and symmetry of the theta distribution is sought about the theta curve upon which the bedding lies (Fig. 4.9A). For some markers such as boudins and concretions it may be reasonable to assume that their pre-flattening attitude was parallel to bedding.

Figure 4.9A shows the $R_f/\emptyset$ pattern such markers should exhibit. As there is no guarantee that the calculated strain by this method is a unique solution, all possible fits should be sought.

Geological Strain Analysis

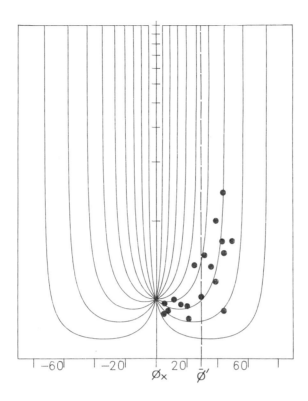

Fig. 4.8 Finding a theta distribution which is symmetrical

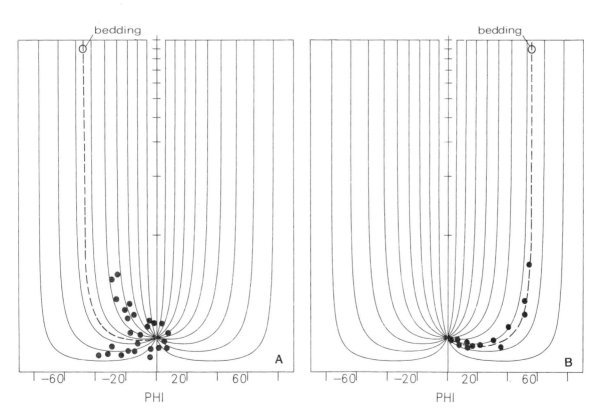

Fig. 4.9 Finding a theta distribution which is symmetrical about bedding (A) and parallel to bedding (B)

If the direction of maximum extension is known on the planar section the search for the marker deformation net which produces a symmetrical theta distribution is made easier. This direction (the X direction of the sectional strain ellipse) will be parallel to the trace of the XY plane of the strain ellipsoid if the section plane is a principal plane of the strain ellipsoid or if the latter is of oblate shape (X = Y > Z). The assumption is frequently made that slaty cleavage is parallel to the XY strain ellipsoid so the cleavage trace under the above specified conditions can be used as the direction of maximum extension. During the comparison of the data points with the marker deformation grids $\phi'_{cleavage}$ is aligned with $\phi = 0$ on the grids so that the 'destraining' takes place about a line corresponding to the cleavage trace (Fig. 4.10).

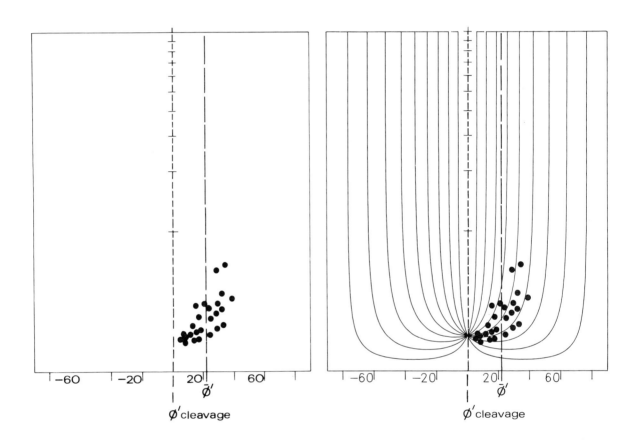

Fig. 4.10 Fitting marker deformation grids when X direction of strain ellipse is known

4.7 OTHER CURVE-FITTING CRITERIA

Assumptions concerning the nature of pre-tectonic markers' orientation form the basis of the curve-fitting procedures described above. Alternative fitting criteria could be adopted which assume specific characteristics for the initial shapes of the markers. Harvey and Ferguson (1981) for instance have used a minimization of mean R_i criterion for recognizing the restored condition. Another would be to restore the markers until their aspect ratios show a frequency distribution similar to that shown by equivalent undeformed material. Alternative strain determination methods using the marker deformation grids could be devised, mutatis mutandis.

5
Extensions of the $R_f/\emptyset$ Method

5.1 INTRODUCTION

The basic method described in the previous chapters is based on the assumption of passive behaviour of the inclusions. This implies that the deformation is homogeneous on the scale of the inclusions and their outlines deform as ellipses inscribed on the deforming material.

There is ample evidence however that many deformed rocks containing inclusions deform in a manner which violates the above assumptions. Observations on the scale of the inclusions frequently reveal heterogeneous strain patterns, often apparently induced by the presence of the inclusions themselves. Recently, methods have been suggested that are extensions of the basic method in that they attempt to take account of inclusions which are not simple markers but are physical inhomogeneities in the rock material.

5.2 VISCOSITY CONTRAST BETWEEN THE INCLUSIONS AND MATRIX

The heterogeneous strains set up in the matrix around a stiffer inclusion lead to the development of a number of structures which are diagnostic of the presence of a competence contrast. These structures are summarised in Figure 5.1 and are discussed by a number of authors (Ramsay 1967, p. 221-226, Ghosh and Sengupta 1973, Strömgård 1973, Ghosh and Ramberg 1976, Shimamoto 1975). These include the swerving of foliation directions and pressure shadows in the matrix.

Such non-passive response on behalf of the inclusion would seem to invalidate the $R_f/\emptyset$ method as originally conceived, because the strain recorded by the inclusions would be different to that suffered by the whole rock. However Gay (1968) pointed out that although competence contrasts between matrix and inclusions may be substantial, their effect would be reduced when the inclusions are densely packed. Intuitively this seems acceptable since if the inclusions form volumetrically a dominant component of the rock, then their average strain cannot differ too much from the strain experienced by the whole rock. Gay's notion of effective competence contrast has often been used in defence of the application of simple methods of analysis which neglect the effects of viscosity contrast.

Where elliptical inclusions are widely-spaced proper account of competence contrasts will need to be taken. Gay (1968), Bilby and others (1975) and Cobbold and Gapais (1983) have considered the problem of the deformation of ellipsoidal viscous inhomogeneities. The work suggests the change of the two dimensional shape of an inclusion seen as a planar section may be a function of the three dimensional shape and orientation of the inclusion and hence that strain analysis will require a three dimensional approach. Furthermore, unlike the passive inclusion case, the deformation of

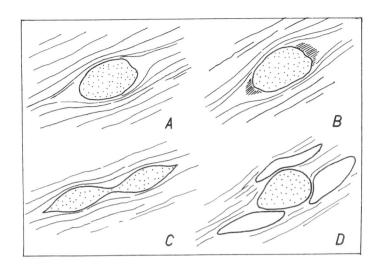

Fig. 5.1 Evidence of competence contrast: deflection of foliation trajectories (A), pressure shadows (B), pinch-and-swell (C) and "interference" between neighbouring inclusions (D).

the viscous inhomogeneity will depend also on the strain history. For example, modification of the inclusion shape will be different for pure shear than for simple shear.

In an attempt to extend the $R_f/\emptyset$ method to include the effects of viscosity contrasts, Lisle and others (1983) have applied the theory of the deformation of elliptical viscous inhomogeneities put forward by Bilby and Kolbuszewski (1977). Although the theory is relevant to markers which are elliptical cross sections through elliptical cylinders, Lisle and others (1977) have applied it as the basis of an approximate method for dealing with sections through ellipsoidal objects.

The equations presented by Bilby and Kolbuszewski (1977), describing the pure shear deformation of viscous elliptical cylinder in terms of its orientation, its axial ratio and the viscosity ratio between cylinder and matrix are:

$$\ln Rs = 2 \int_{Ri}^{Rf} \frac{J(R)G(R)}{2R(R+1)^2 [G^2(R) - G^2(Ri)\sin^2 2\theta]^{\frac{1}{2}}} dR \qquad (5.1)$$

and:

$$\sin 2\emptyset = \frac{G(Ri)}{G(Rf)} \sin 2\theta \qquad (5.2)$$

where

$$J(R) = R^2 + 2VR + 1 \qquad (5.3)$$

$$G(R) = \frac{R^2 - 1}{R} \left[\frac{VR^2 + 2R + V}{(R+1)^2} \right]^v \qquad (5.4)$$

and V = viscosity ratio, μ_{pebble}/μ_{rock}. These equations allow the calculation of final orientation and axial ratio of an elliptical inclusion whose initial orientation and axial ratio is known, together with the shape and orientation of the strain ellipse. They can be used to calculate marker deformation grids for different R_s values and V values. Clearly the number of such grids required to span a reasonable range of V and R_s is enormous and no attempt is made to present them in this book. Instead, pure shear deformation paths are given for various viscosity contrasts. These are presented in Chapter 7 where it is explained how they can be used to generate marker deformation grids for any required strain and viscosity contrast.

Once marker deformation grids have been produced they can be used in the manner set out in Chapter 4 to determine whole rock strains and viscosity contrasts from $R_f/\emptyset$ data.

Application of this method to conglomerates (Lisle and others 1983) has revealed that this technique sometimes lacks precision, in that several combinations of V and R_s can successfully de-strain the markers to the assumed initial state. This can be overcome by placing further constraints on V and R by either separately analysing inclusions in a rock which belong to more than one competence group (Lisle and others 1983) or by obtaining an independent estimate of the whole rock strain using a strain analysis technique which is resistant to the effects of marker/matrix competence contrast such as marker separation method of Fry (1979).

It should be emphasised that the graphs showing deformation paths in Chapter 7 are valid only for a pure shear deformation history. Unlike the passive marker model, the type of deformation history has an influence on the deformed shapes and orientations of viscous inhomogeneities. The marker deformation paths will have a different shape if the strain history is for example, that of simple shear (Bilby and Kolbuszewski 1977).

5.3 DEFORMATION BY PRESSURE SOLUTION

Another type of inhomogeneous strain pattern results from the operation of pressure solution along planar zones in a rock. Grains positioned along such zones or channel-ways (Williams 1972) will change shape by the selective removal of material from parts of the grain adjacent to the channel way. The result of this process is the production of a fabric which can be analyzed using an $R_f/\emptyset$ form of representation. $R_f/\emptyset$ plots from such rocks are distinctive and are characterized by showing a higher fluctuation of orientation than expected from the passive marker model for a corresponding R_f distribution (Lisle 1977b). Geometrical modelling of fabric production by pressure solution has been carried out by Onasch (1984). The model predicts $R_f/\emptyset$ patterns which differ significantly from those produced by the homogeneous strain of passive ellipses.

5.4 DEFORMATION OF ANISOTROPIC MARKERS

Other curious types of $R_f/\emptyset$ pattern have been observed in rocks containing markers which are mechanically anisotropic. Marjoribanks (1976), Carreras and others (1977), Bouchez (1977) and Mancktelow (1981) all present $R_f/\emptyset$ patterns of quartz grain shapes in which the range of R_f values are abnormally high. Mancktelow (1981) attributes this pattern to the influence of anisotropy which leads to the large discrepancy between the shapes adopted by grains which are favourably- and unfavourably-oriented with respect to the bulk strain axes. In consequence R_f is more markedly dependent on $\emptyset$ than is predicted by the passive model.

5.5 THE $R_f/\emptyset$ METHOD APPLIED TO HETEROGENEOUSLY STRAINED ROCKS

The Ramsay-Dunnet method is designed to analyze rocks where variation in the shape and orientation of elliptical objects results from the homogeneous deformation of initially variably-oriented marker shapes. Holm (1983) has applied the technique to experimentally deformed oolitic limestone where an important contribution to the variability of R_f and $\emptyset$ was ascribed to the heterogeneous strain of passive markers (ooids). Values of strain calculated using the $R_f/\emptyset$ method were in close agreement with the known strain in the deformed samples. This suggests that the $R_f/\emptyset$ method may provide a useful way of "averaging" strains in a heterogeneous strained body. The validity of this application of the $R_f/\emptyset$ method has still to be justified theoretically.

6

The Marker Deformation Grids

This section contains the standard reference curves for R_f/ϕ analysis of passive markers. If corresponding charts are required for the situation involving a viscosity contrast between markers and matrix, the reader is referred to section 5.2 and Chapter 7.

The charts given below correspond to a range of strain ratios which covers the strain magnitude most frequently encountered in deformed rocks. Supplementary deformation grids for intermediate values of strain can be generated using the graphs in Chapter 7. The appropriate graph is the one for a viscosity contrast of 1 (page 82).

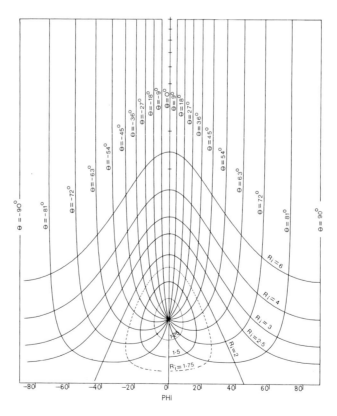

Fig. 6.1 A typical deformation grid with labelled R_i and θ curves

Figure 6.1 shows a typical marker deformation grid on which the R_i and θ curves have been annotated. When $R_s > 4.0$ the curves for $R_i = 1.25$ and $R_i = 1.75$ are omitted and when $R_s > 10.0$ only the $R_s = 2.0$ curve is shown.

The Marker Deformation Grids

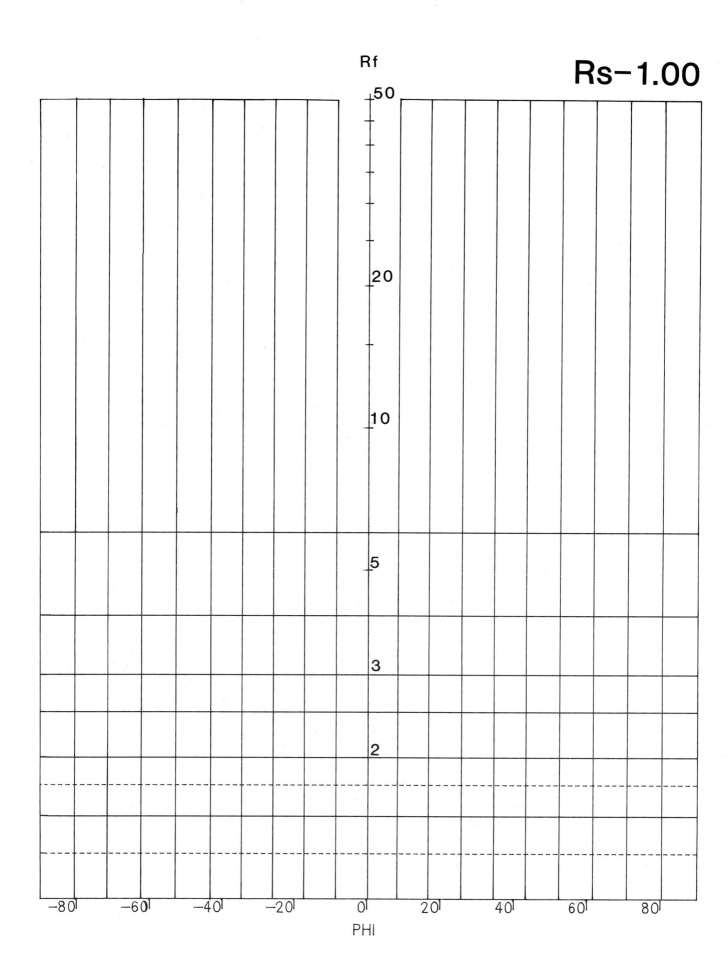

Rs-1.00

Geological Strain Analysis

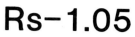

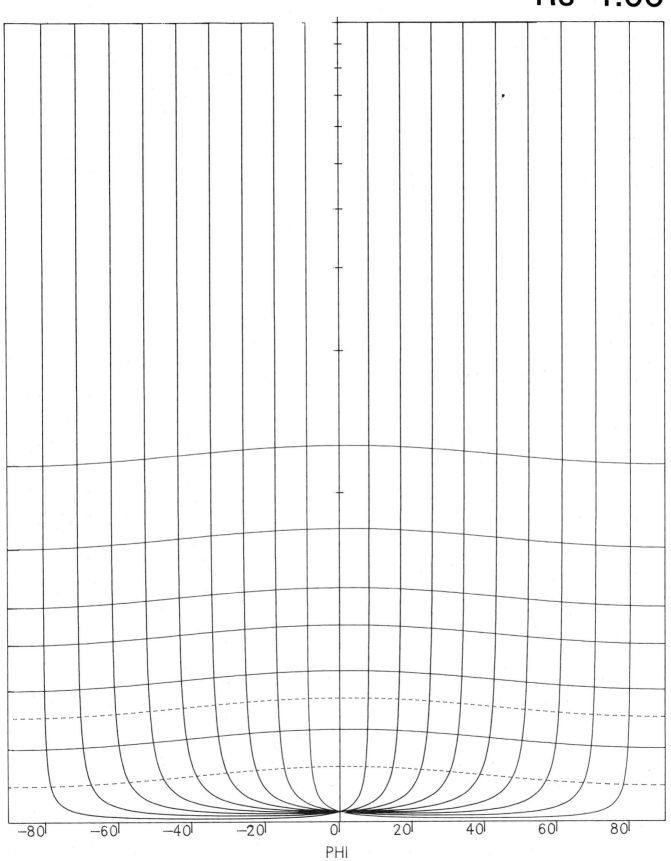

PHI

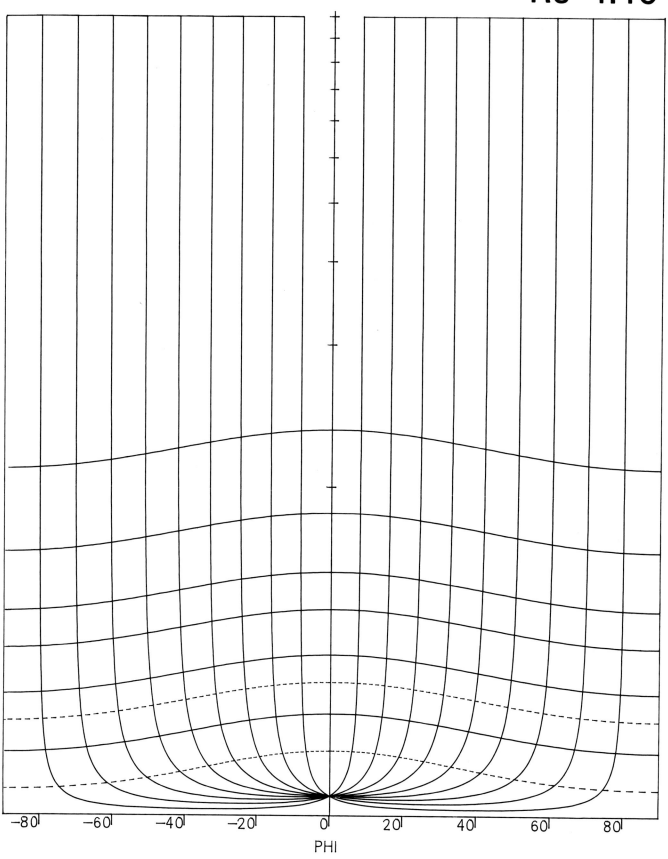

Rs-1.10

Geological Strain Analysis

Rs-1.15

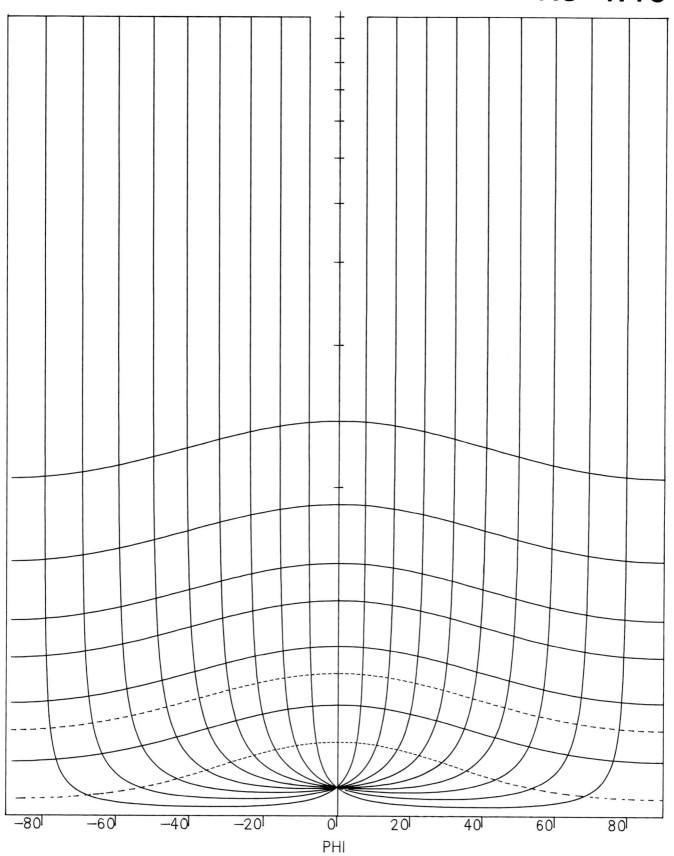

PHI

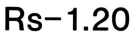

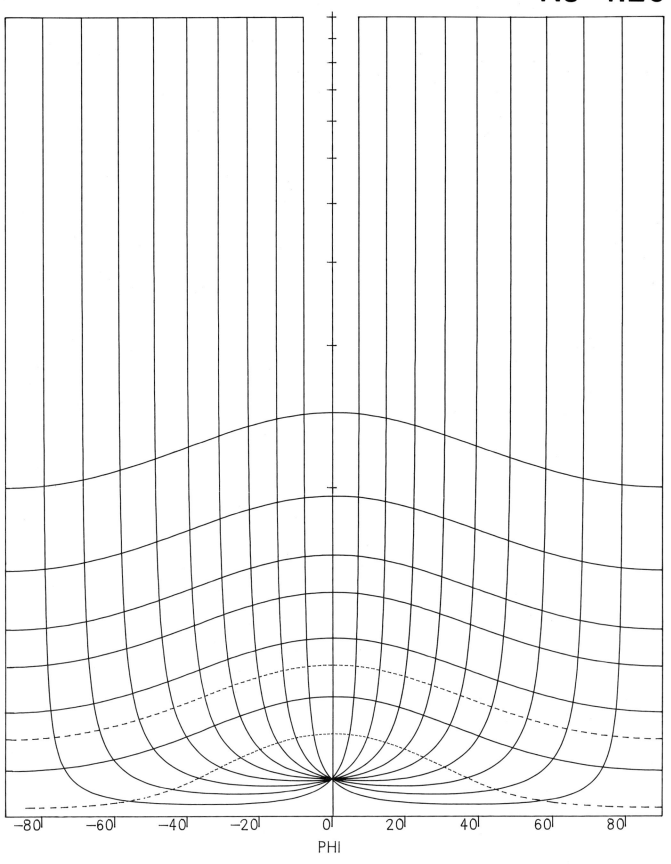

PHI

Rs-1.25

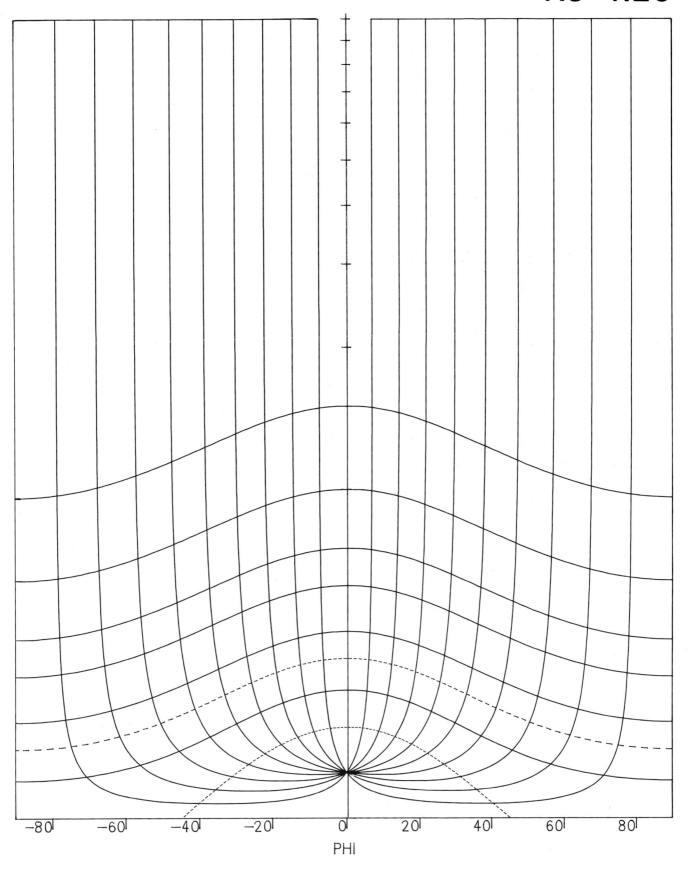

PHI

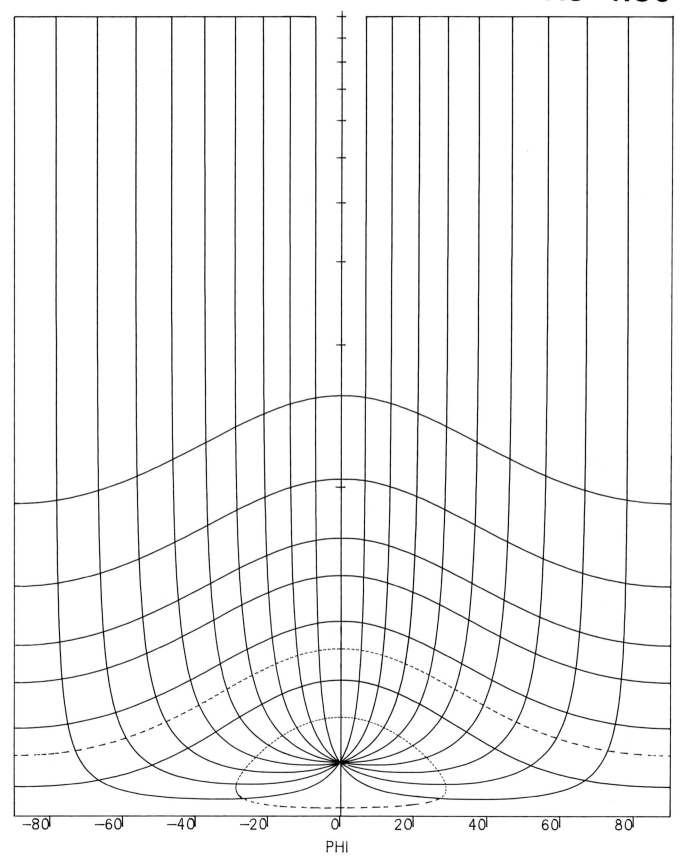

Geological Strain Analysis

Rs-1.35

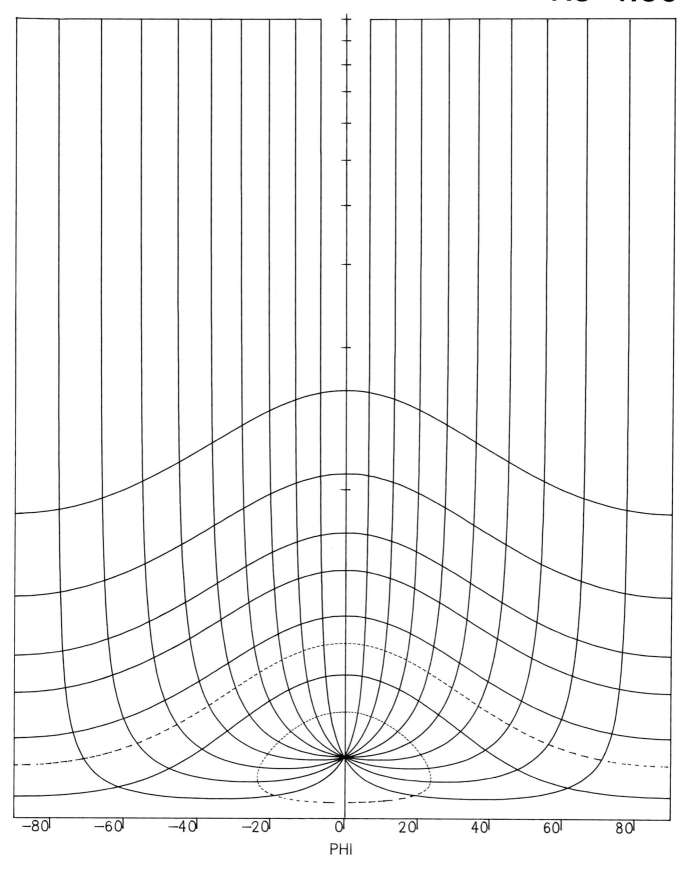

PHI

Rs-1.40

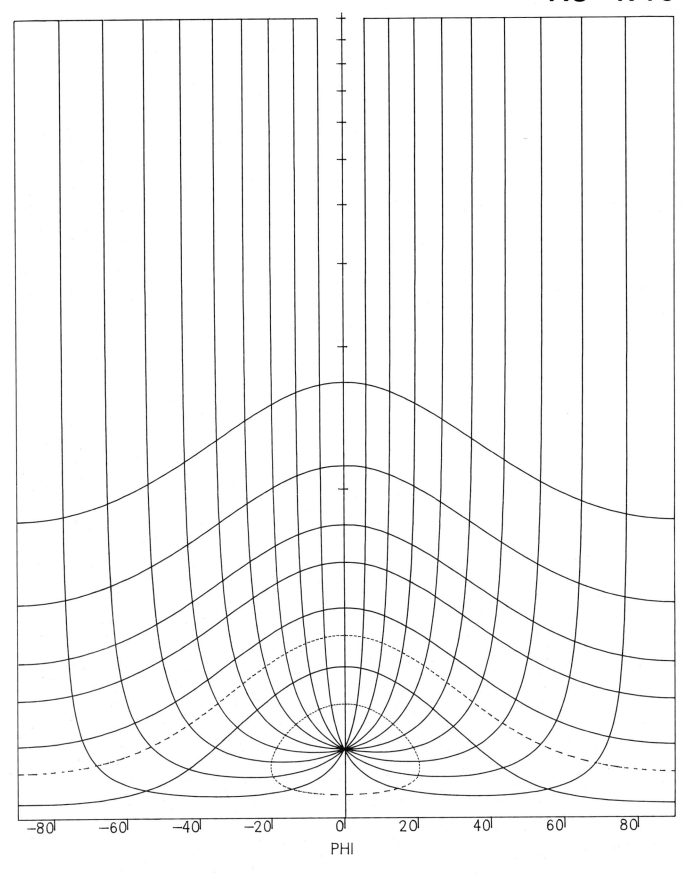

PHI

Geological Strain Analysis

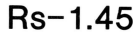

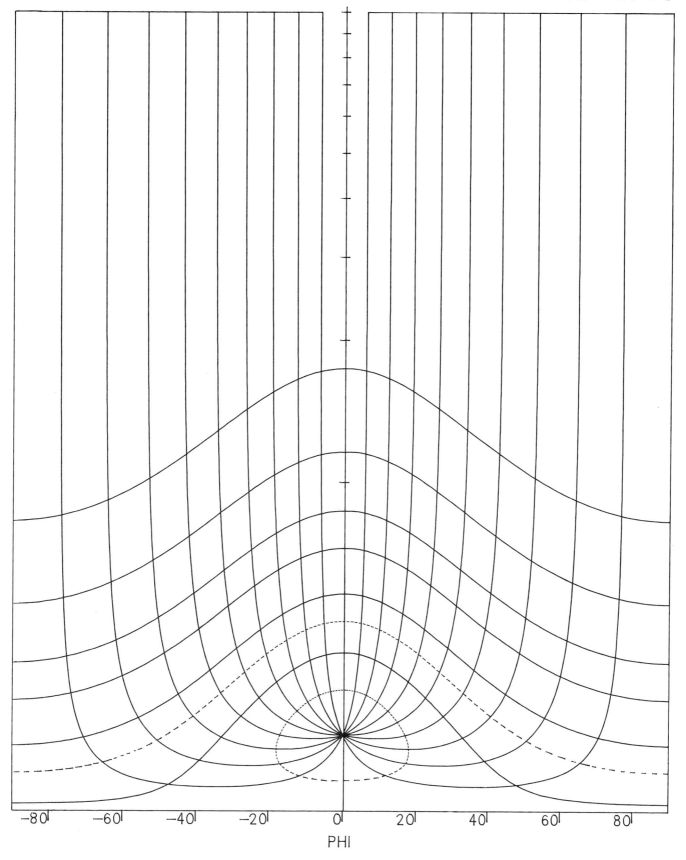

PHI

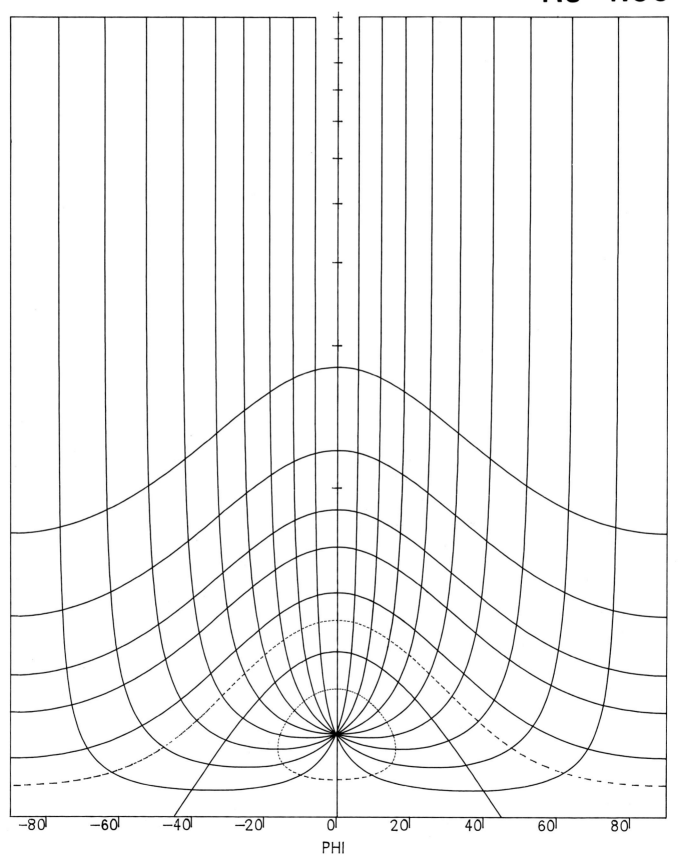

Geological Strain Analysis

Rs-1.60

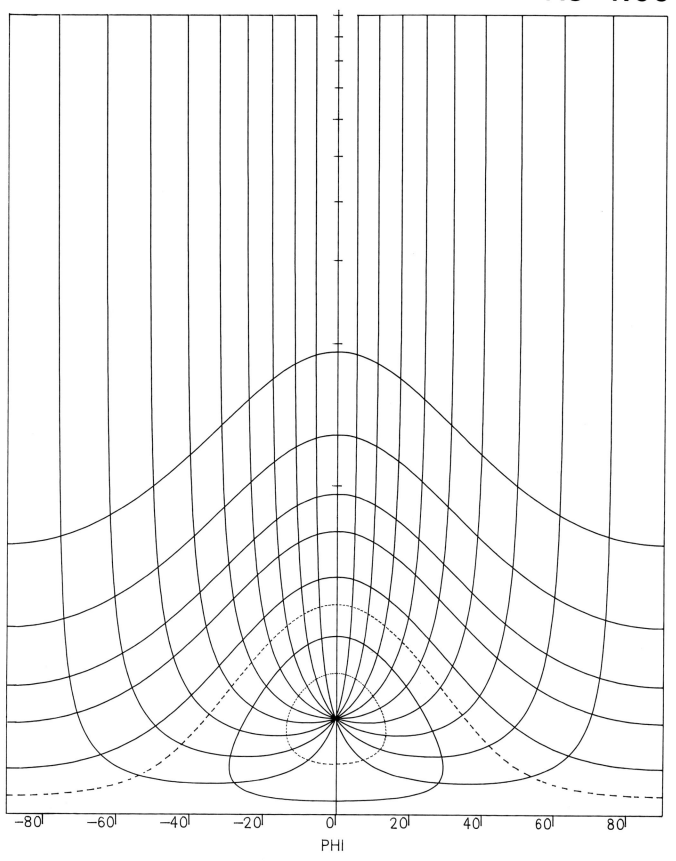

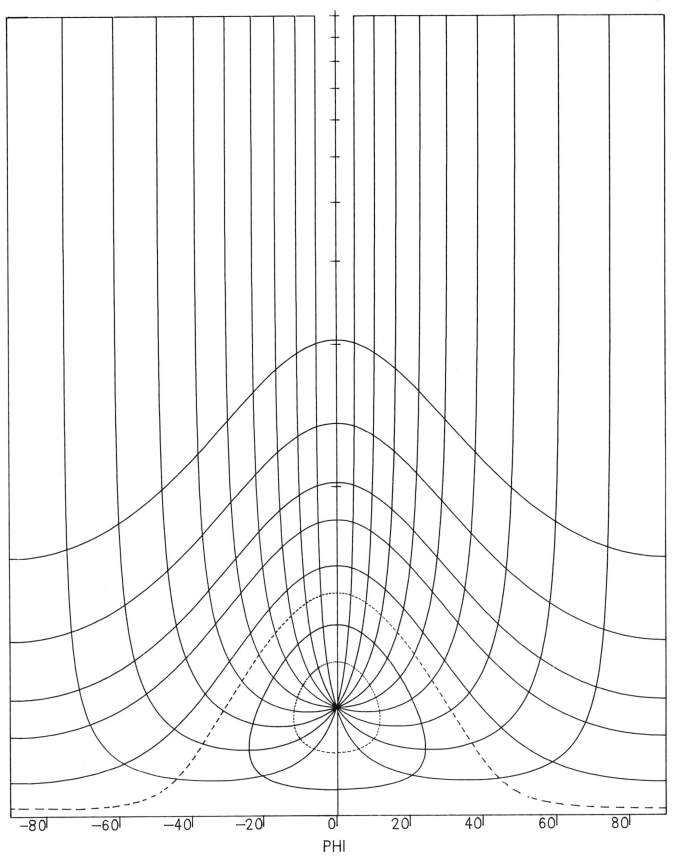

Geological Strain Analysis

Rs-1.80

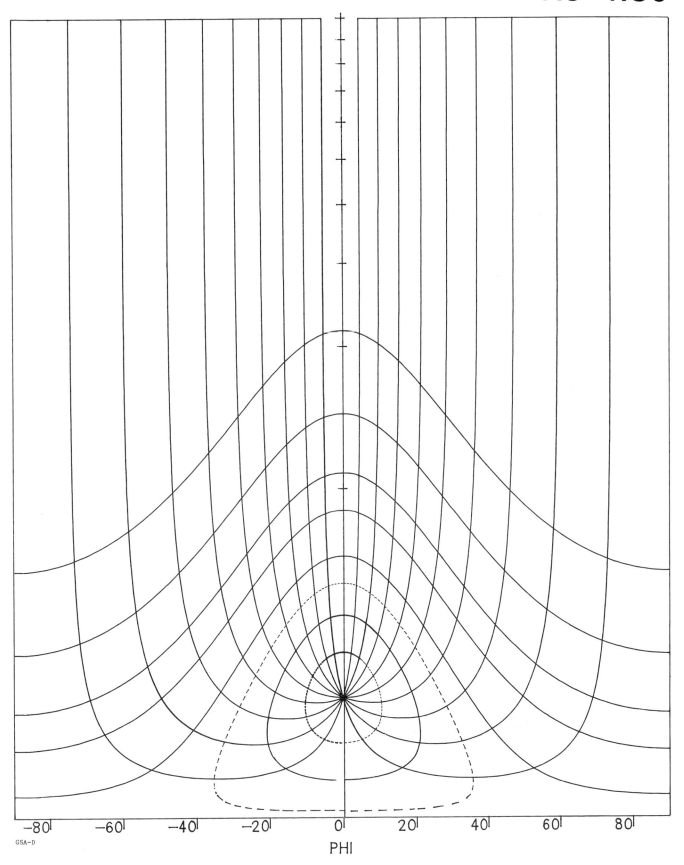

PHI

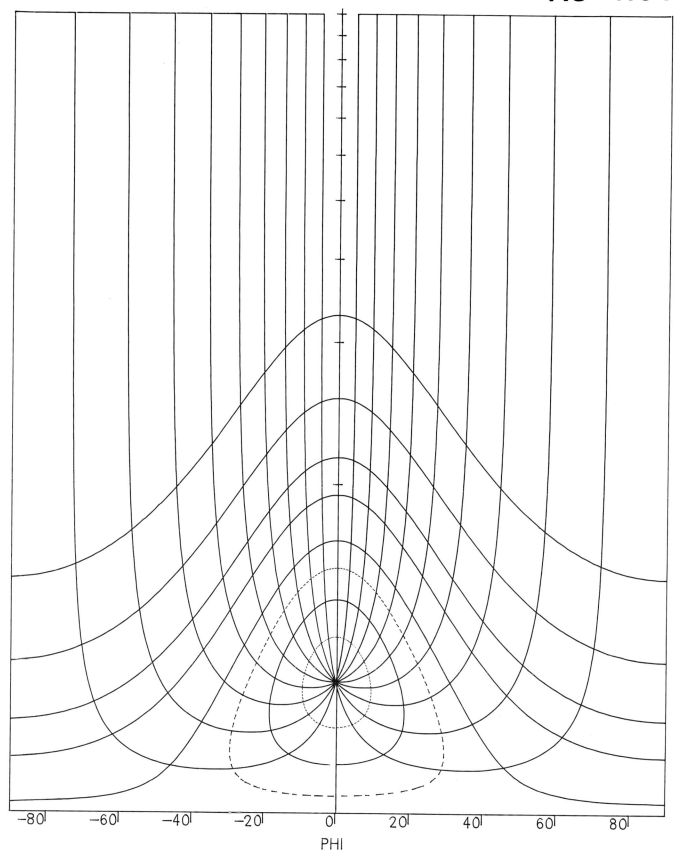

Rs-1.90

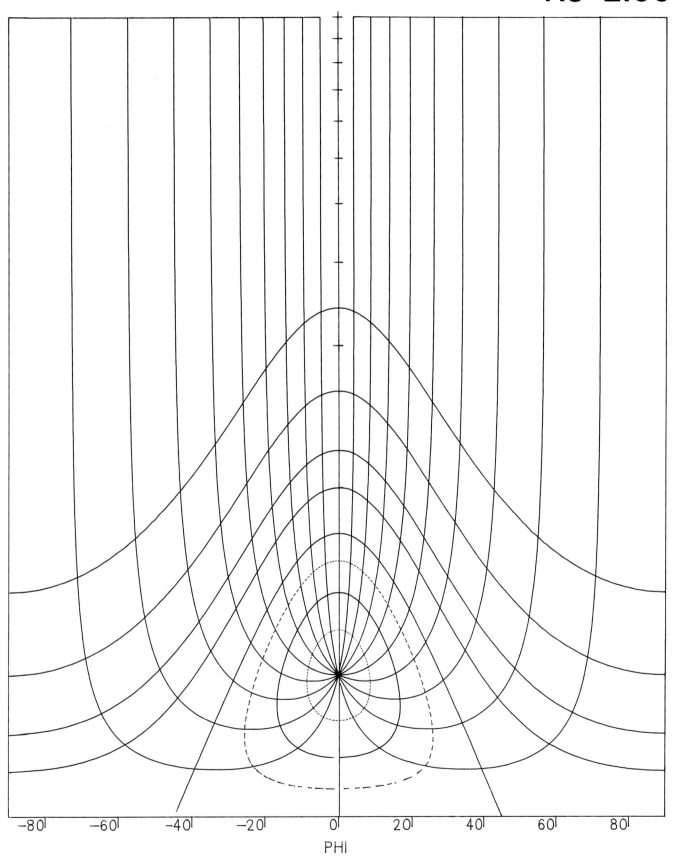
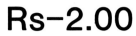

Rs-2.10

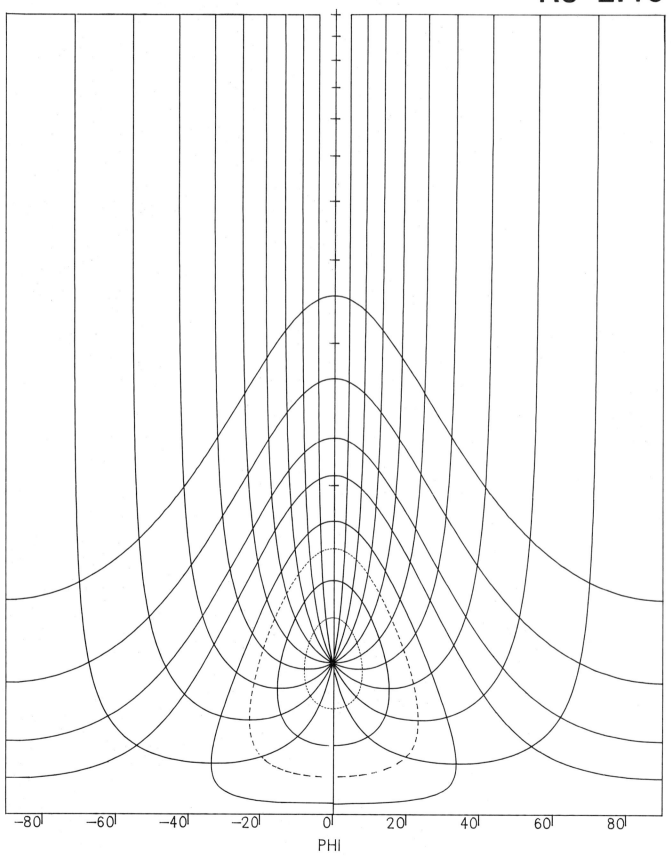

PHI

Geological Strain Analysis

Rs-2.20

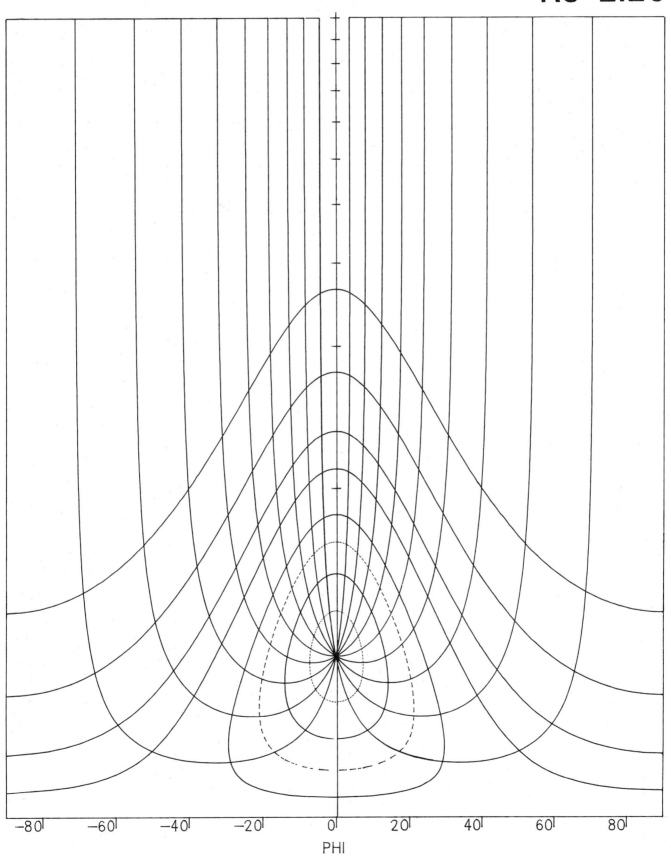

PHI

Rs-2.30

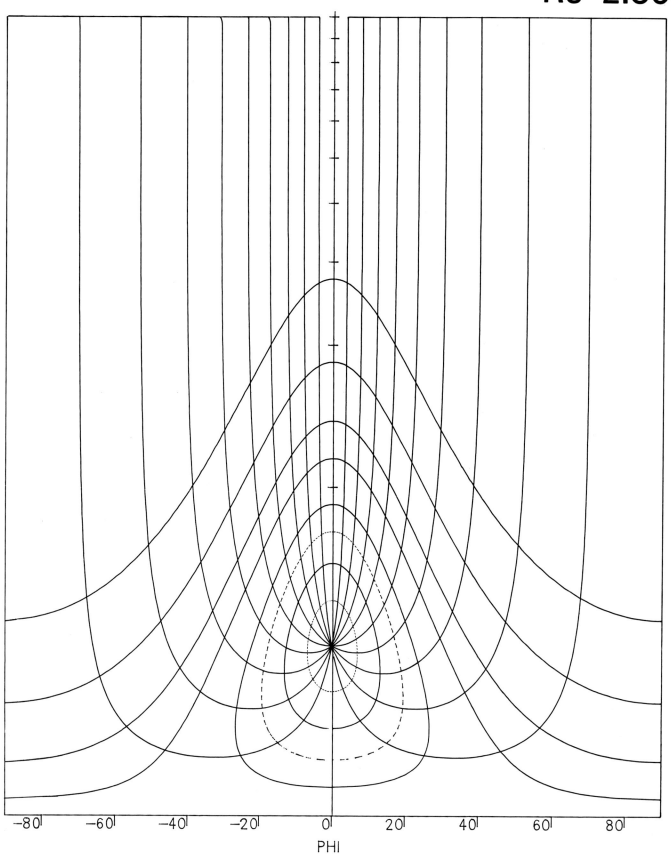

PHI

Geological Strain Analysis

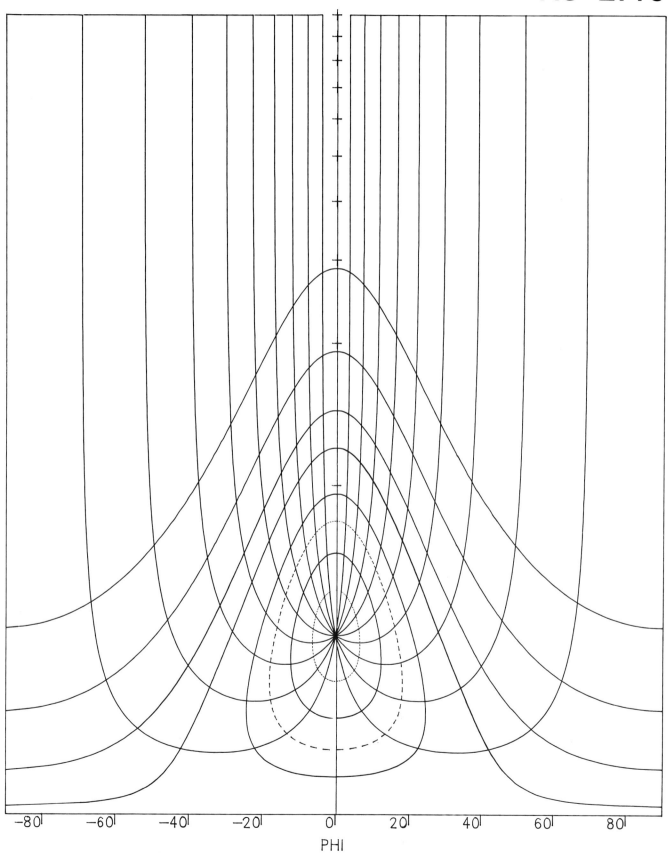

Rs-2.40

PHI

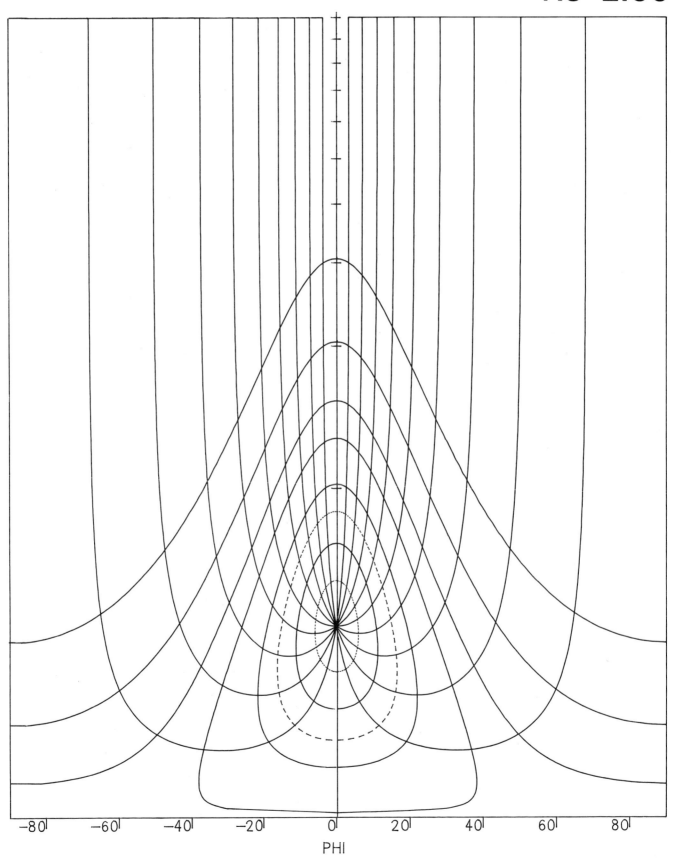

Geological Strain Analysis

Rs-2.70

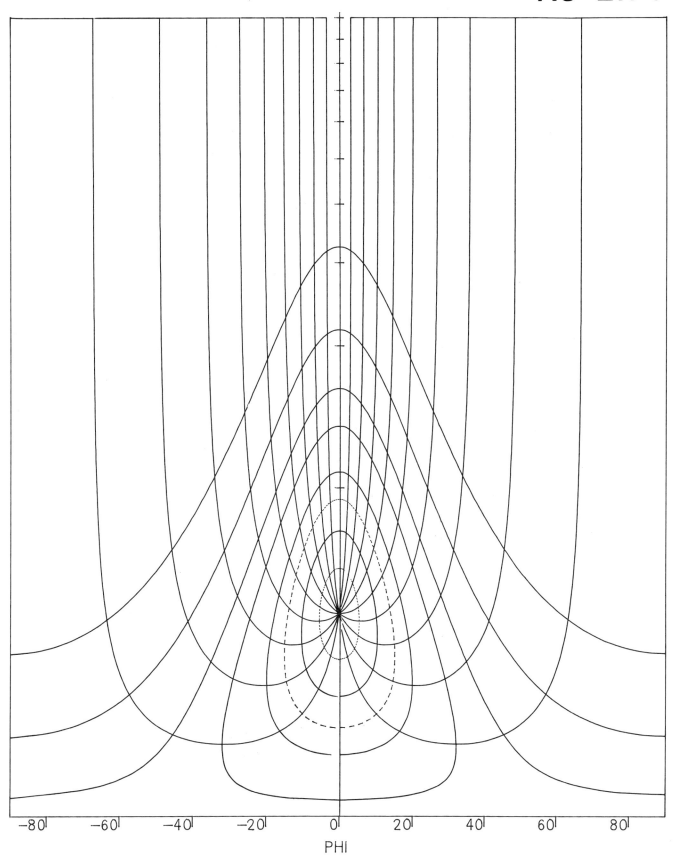

PHI

Rs-2.85

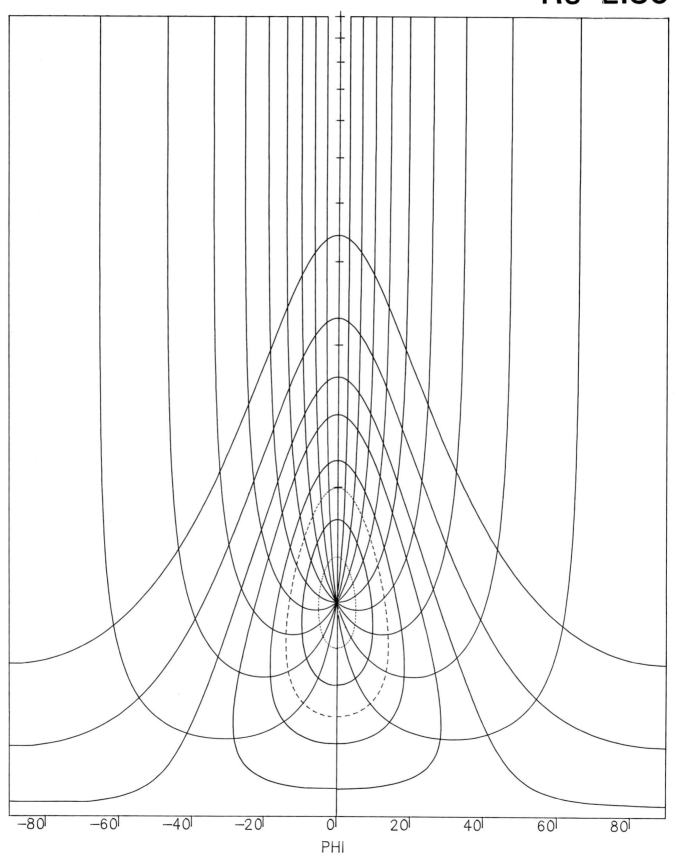

PHI

Geological Strain Analysis

Rs-3.00

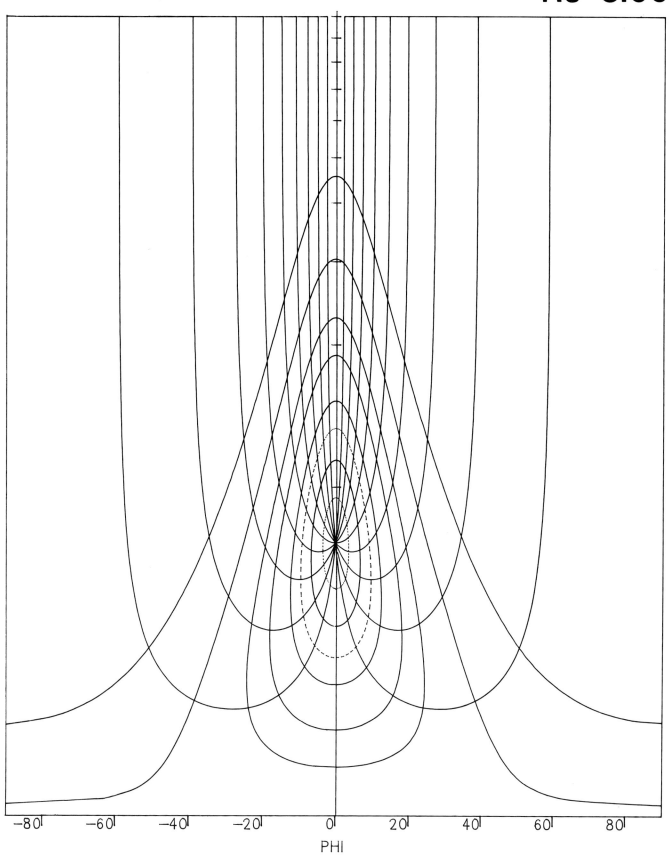

PHI

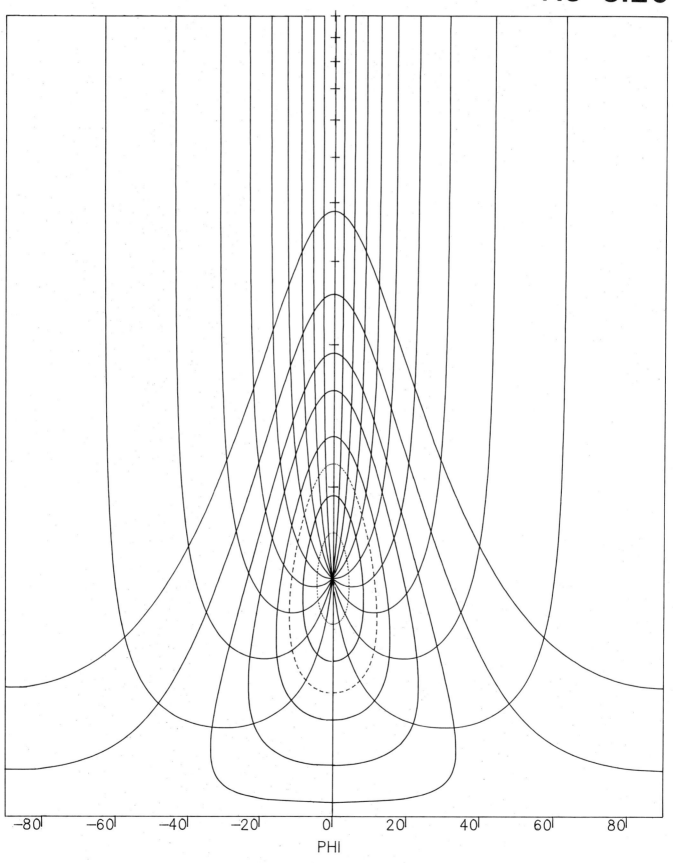

Geological Strain Analysis

Rs-3.40

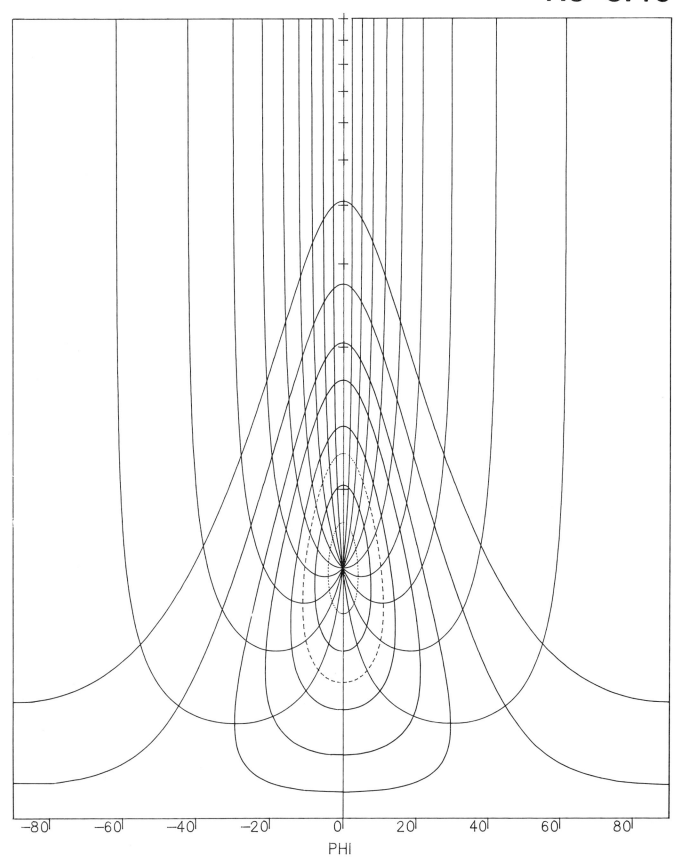

PHI

Rs-3.60

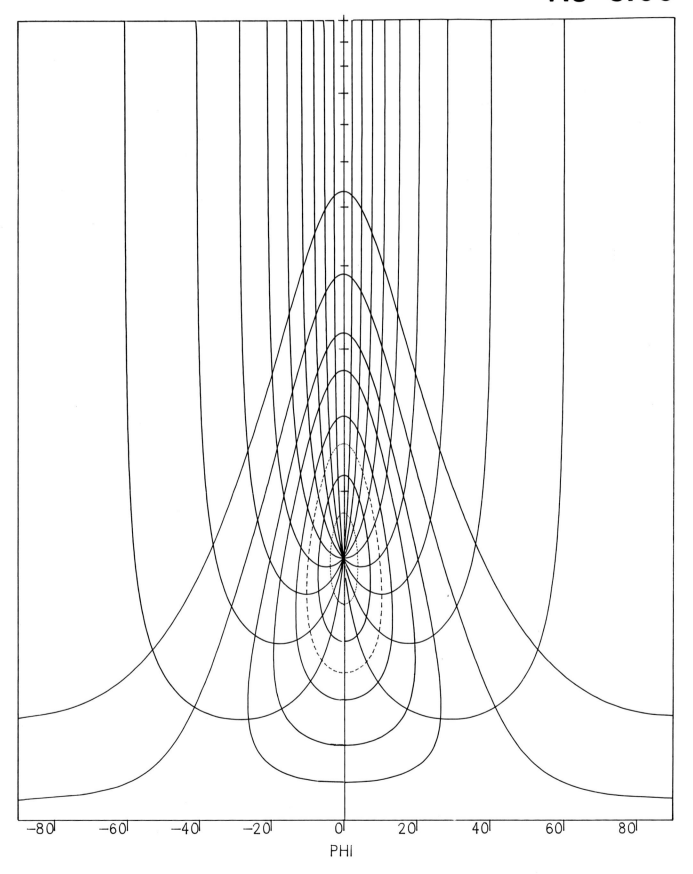

PHI

Rs-3.80

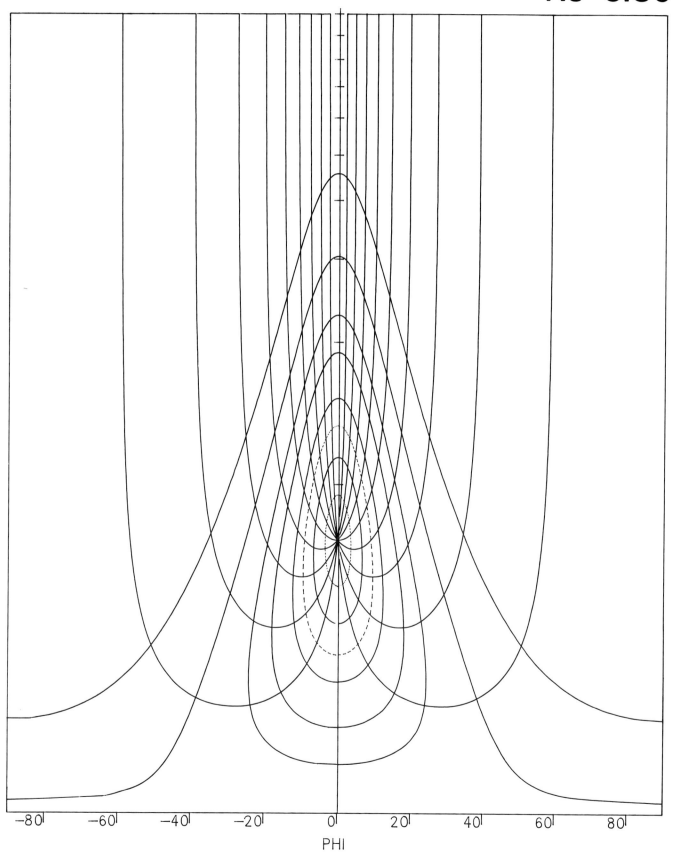

PHI

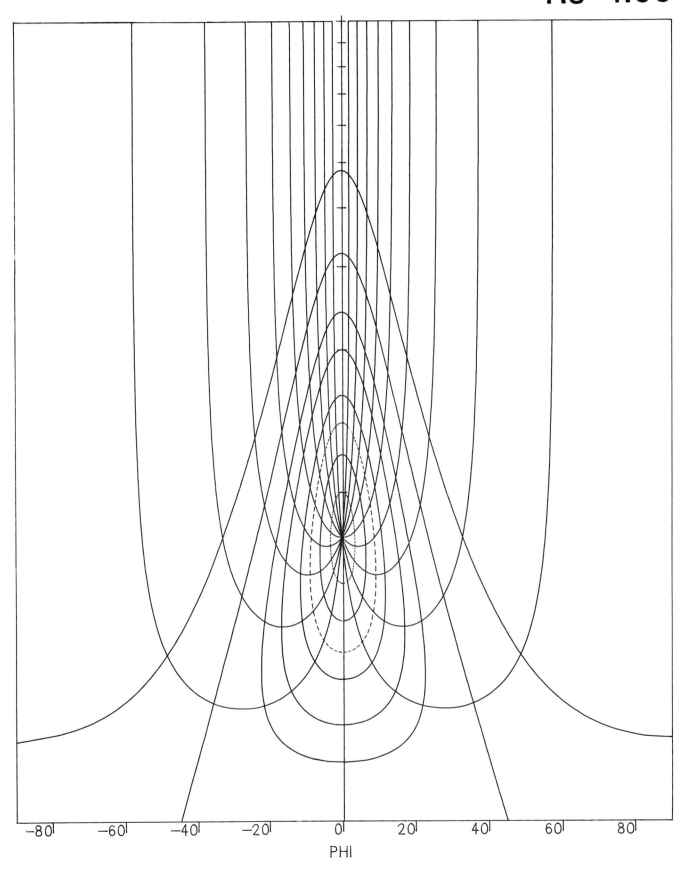

Rs-4.20

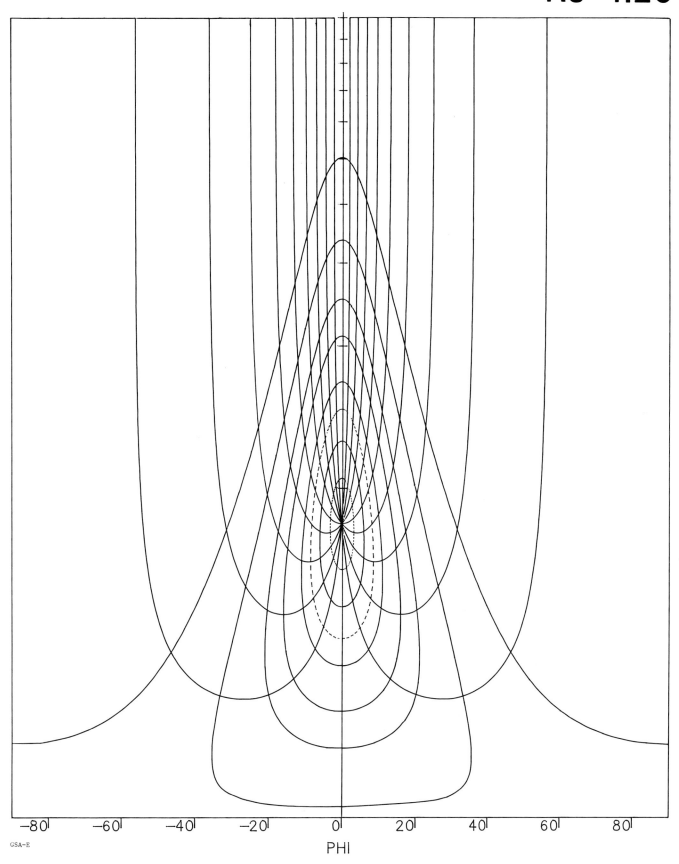

PHI

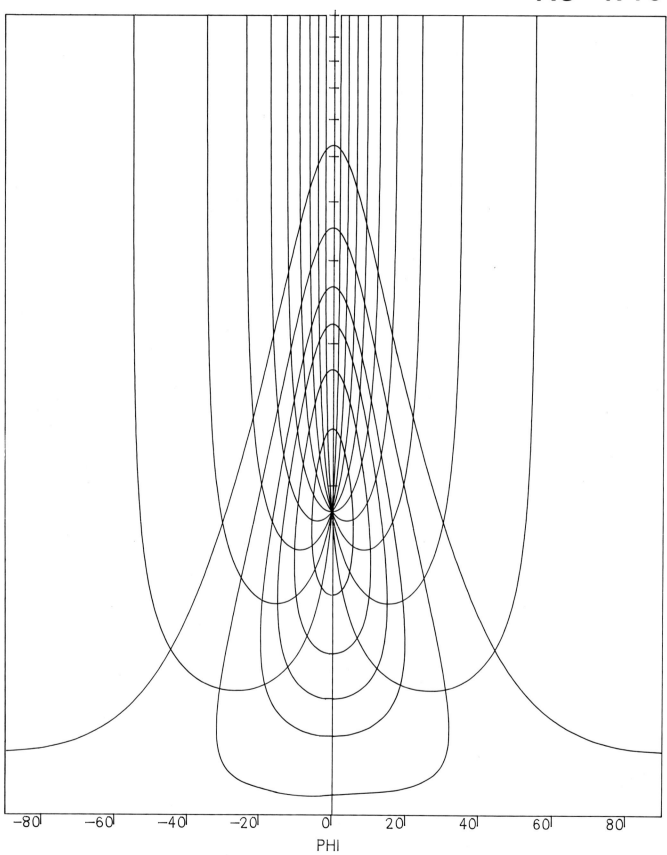

Rs-4.40

Geological Strain Analysis

Rs-4.60

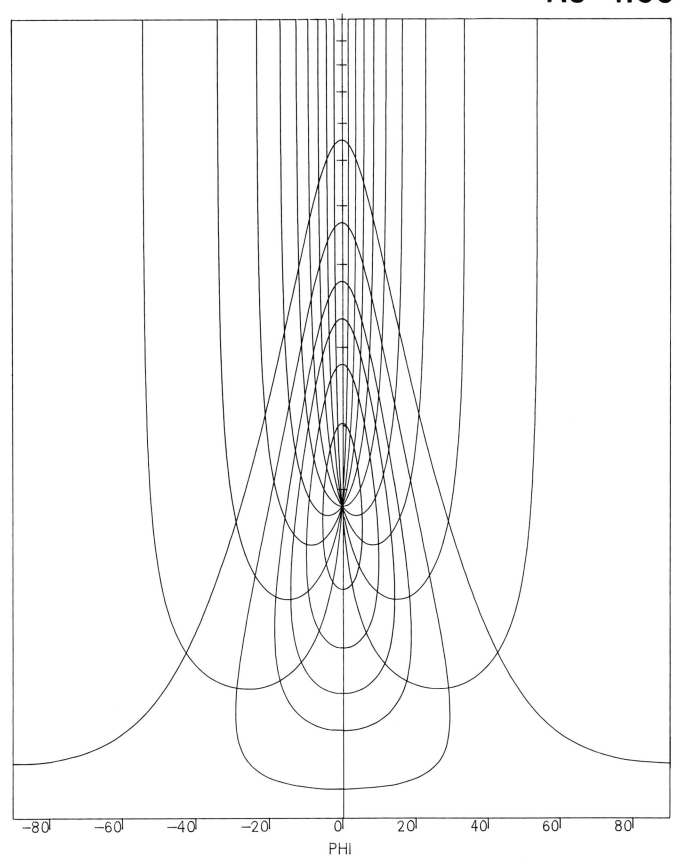

PHI

Rs-4.80

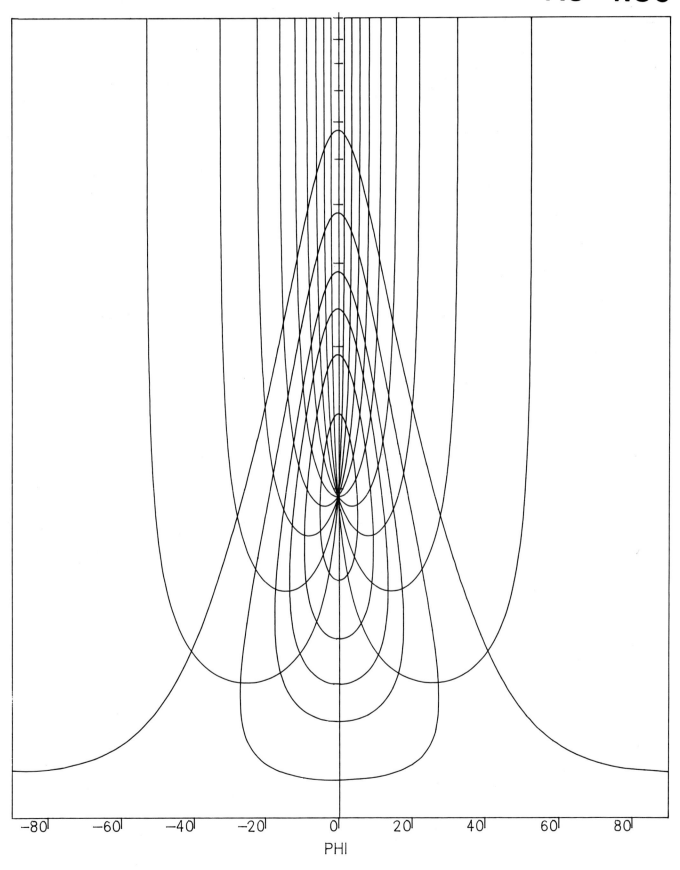

PHI

Geological Strain Analysis

Rs-5.00

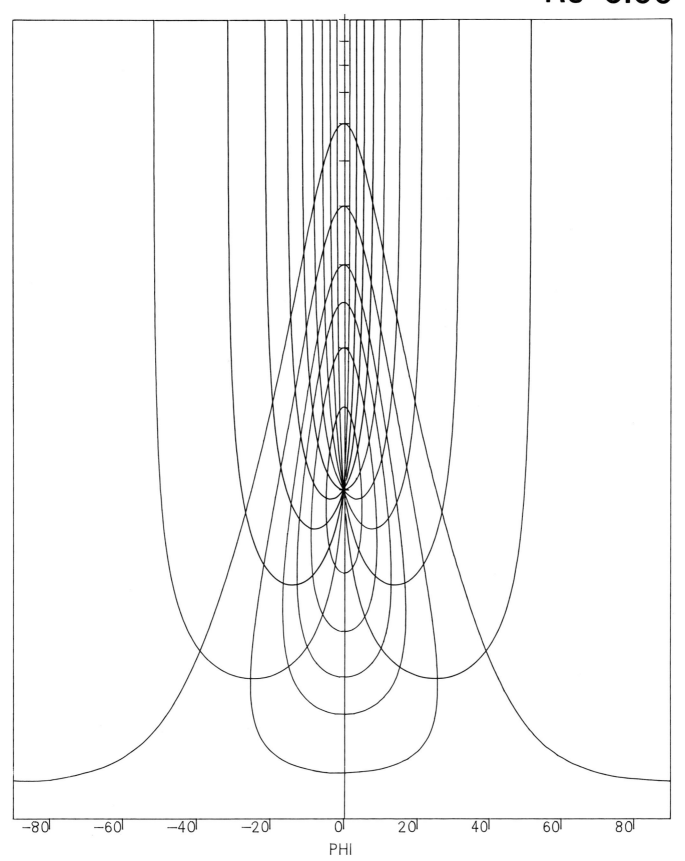

PHI

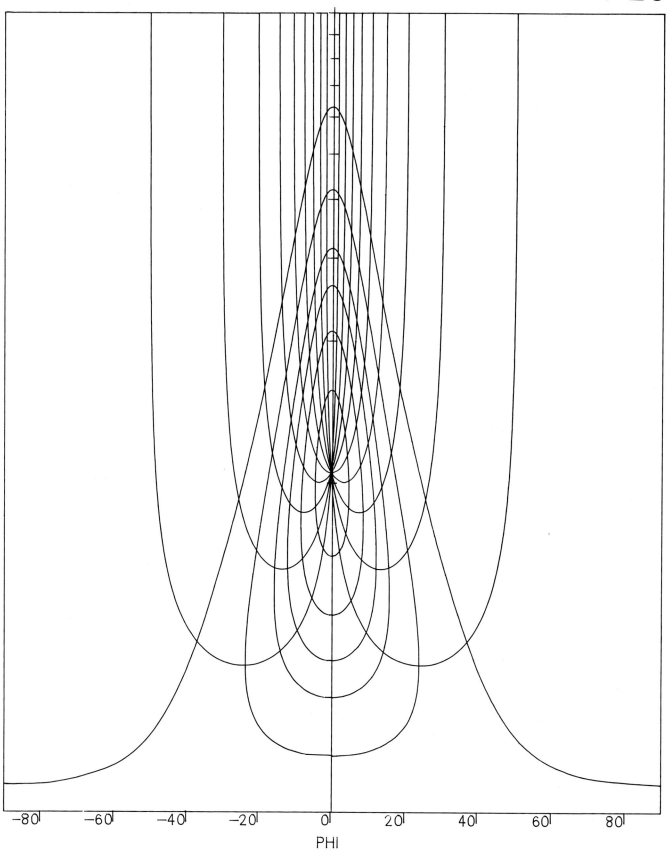

Geological Strain Analysis

Rs-5.50

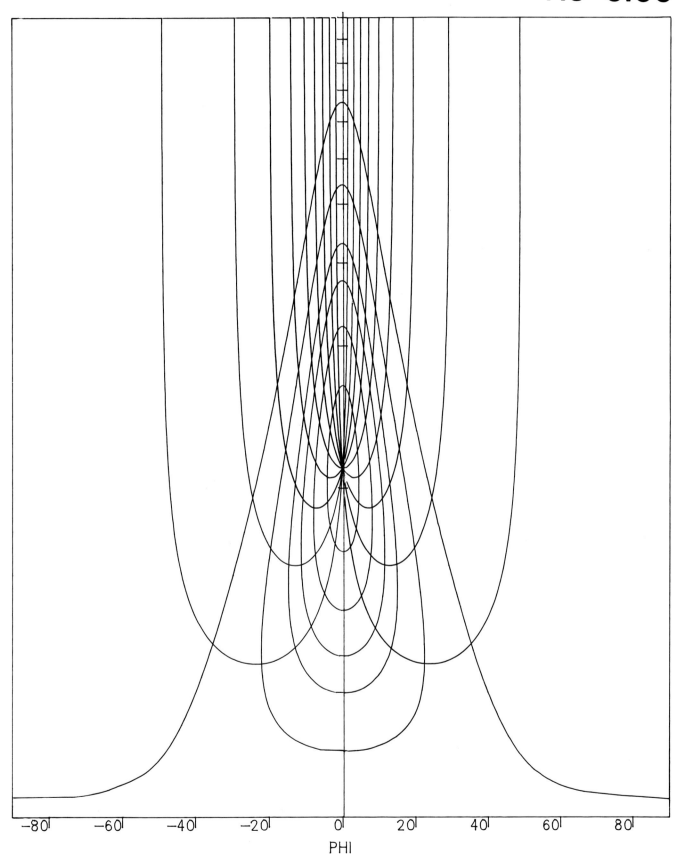

PHI

Rs-5.75

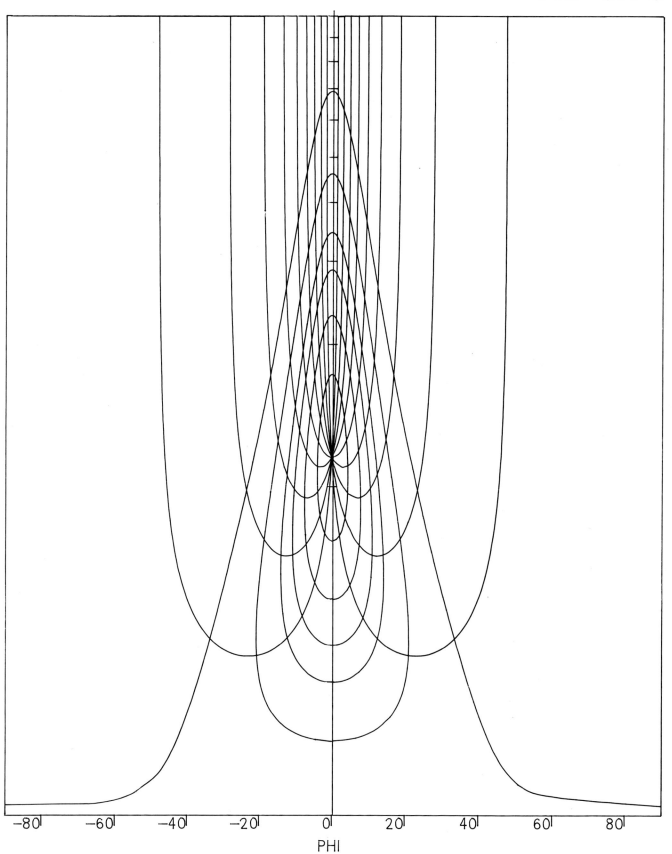

Geological Strain Analysis

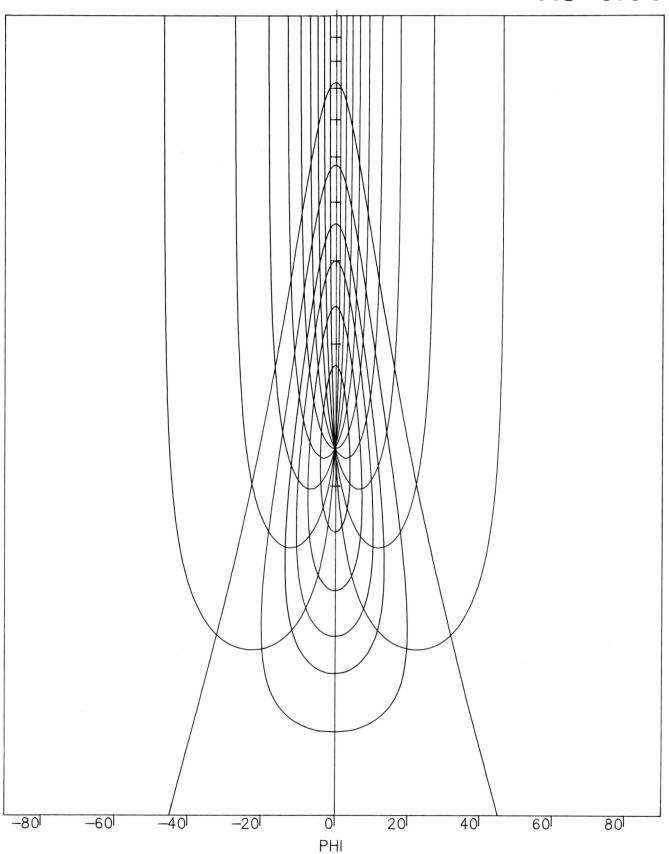

Rs-6.25

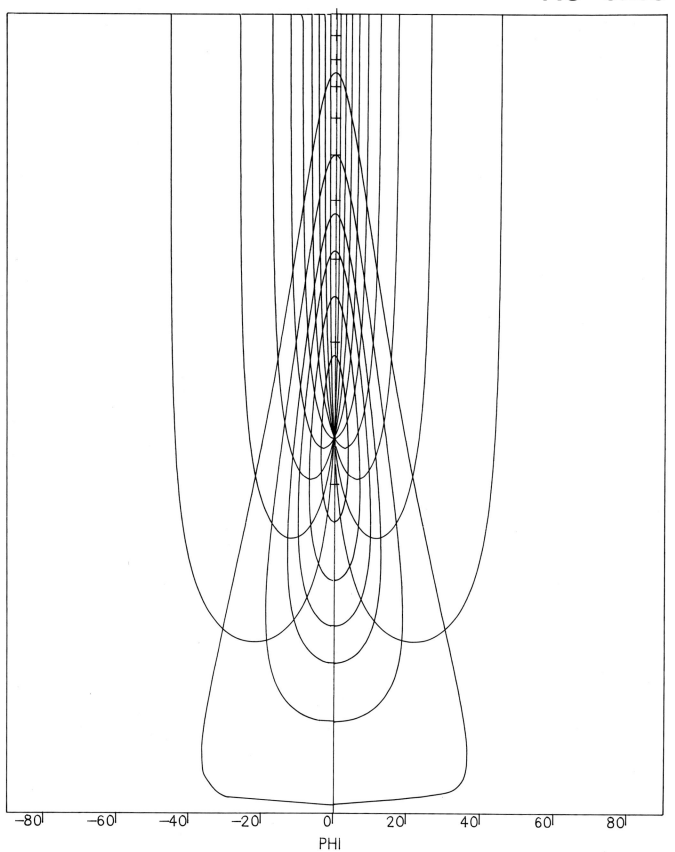

PHI

Rs-6.50

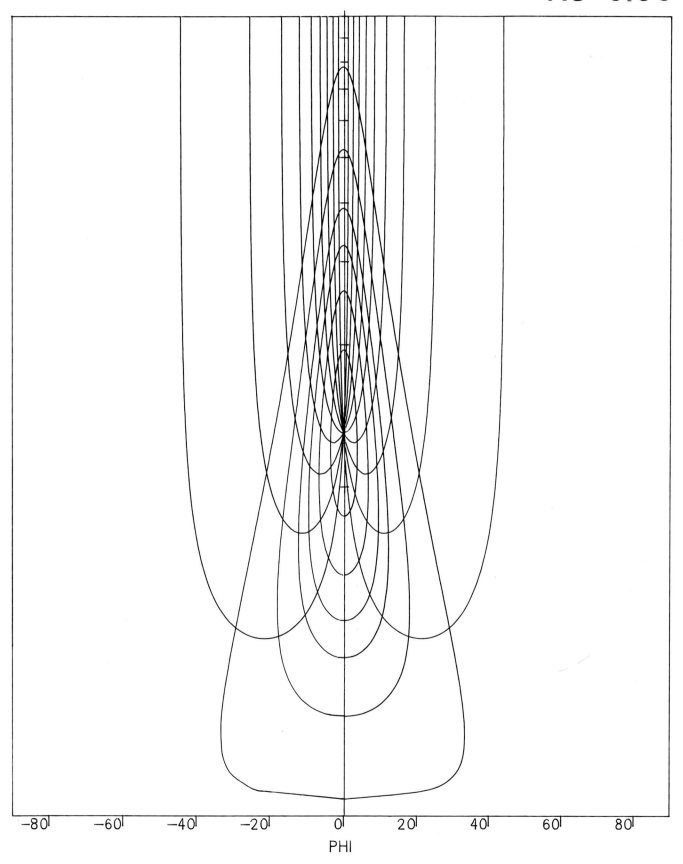

PHI

Rs-6.75

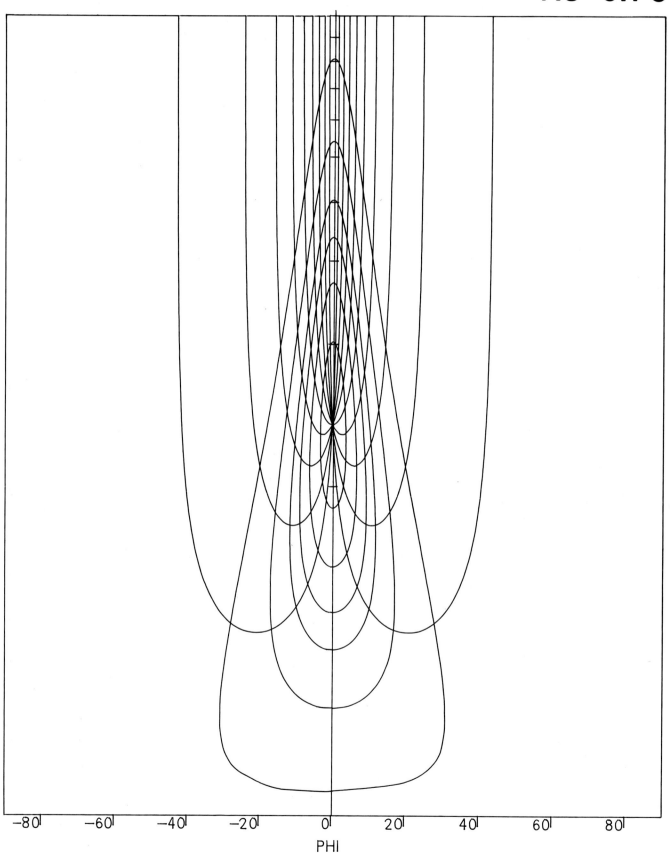

Rs-7.00

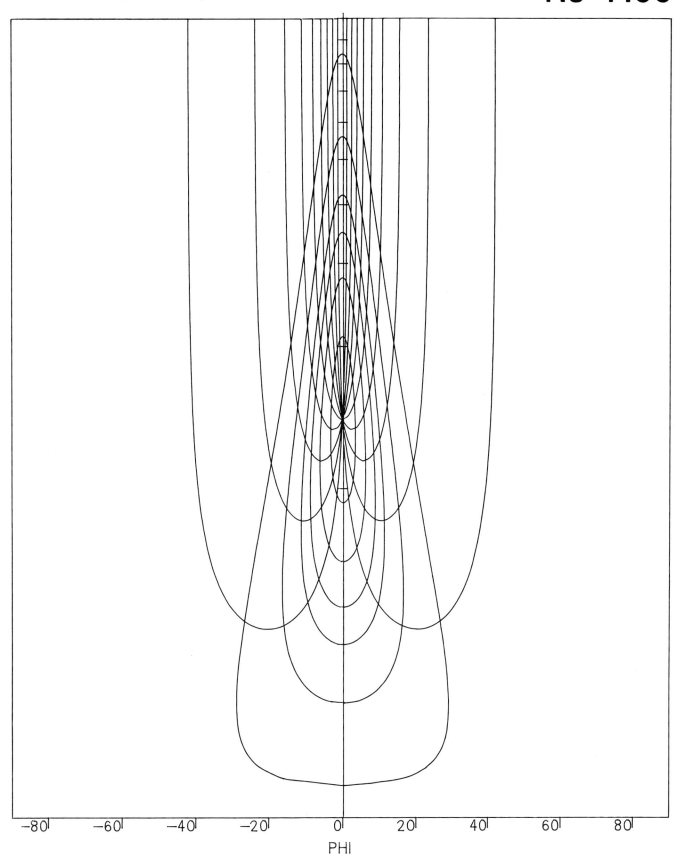

PHI

Rs-7.50

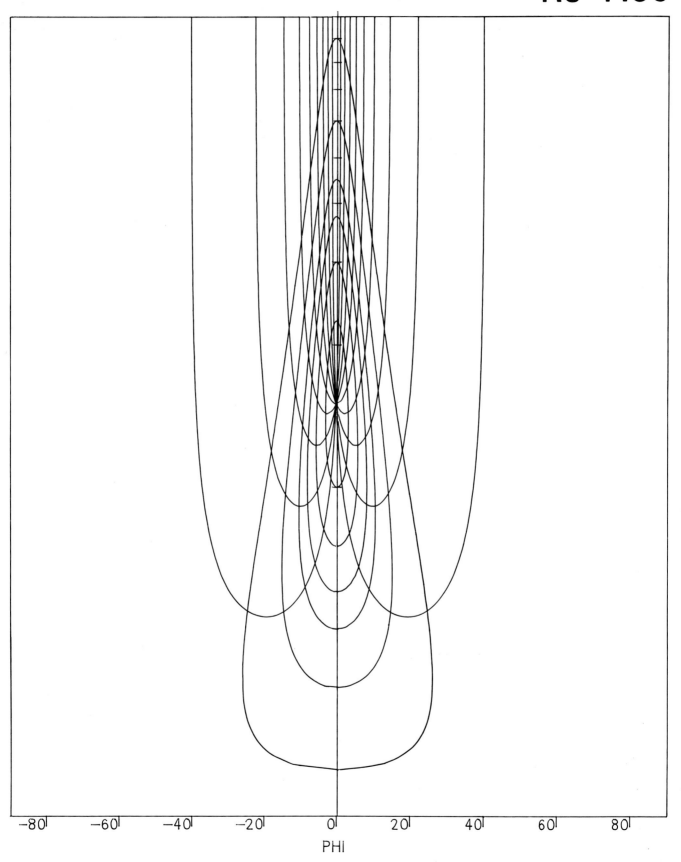

Geological Strain Analysis

Rs-8.00

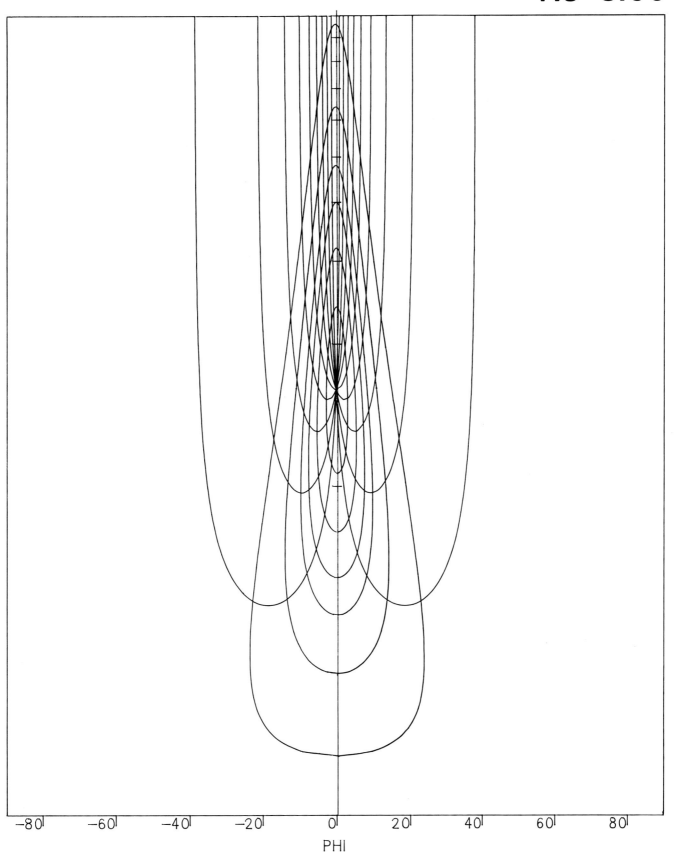

PHI

Rs-8.50

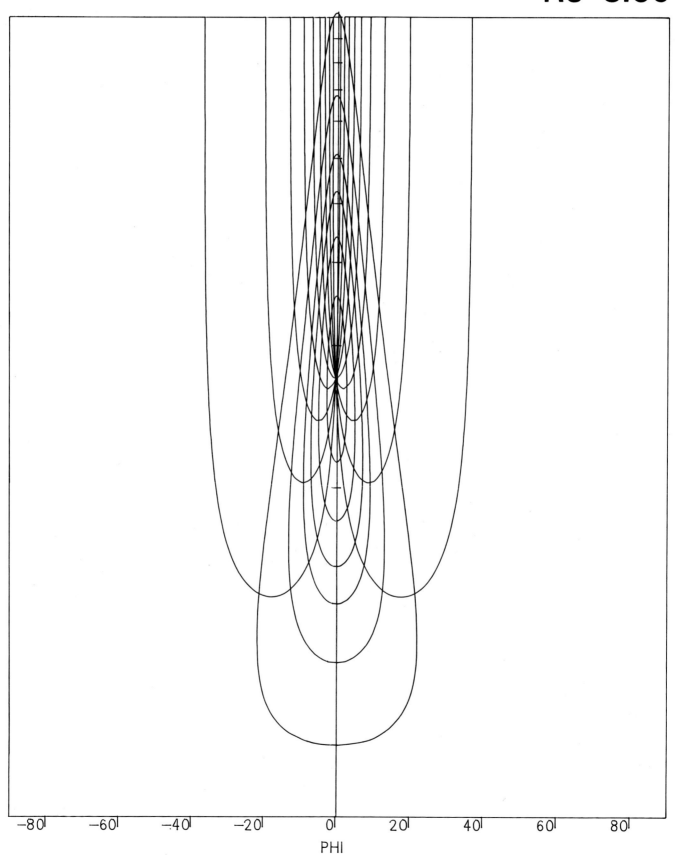

PHI

Geological Strain Analysis

Rs-9.00

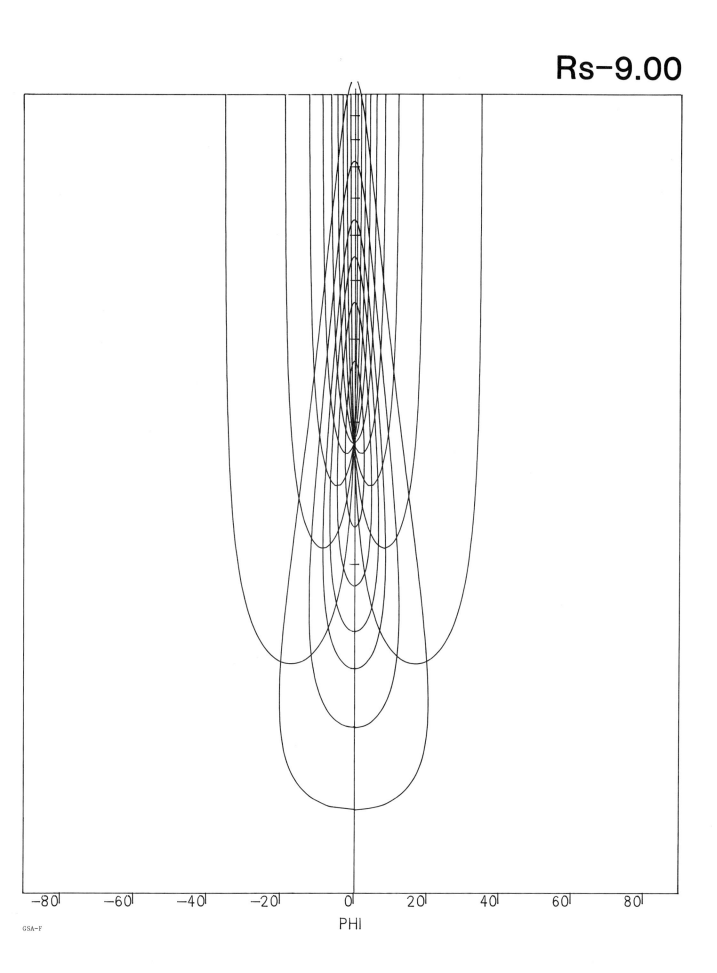

PHI

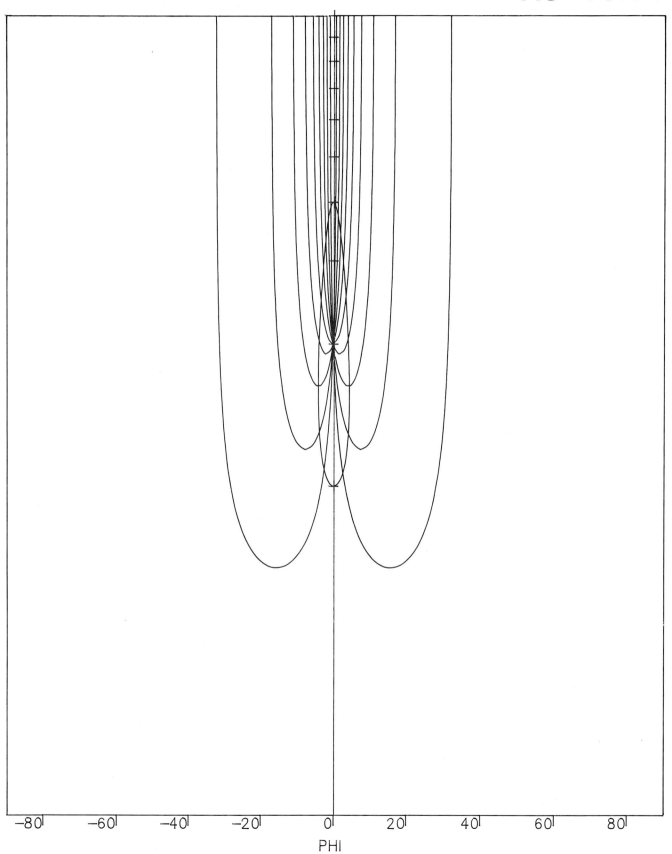

Geological Strain Analysis

Rs-12.00

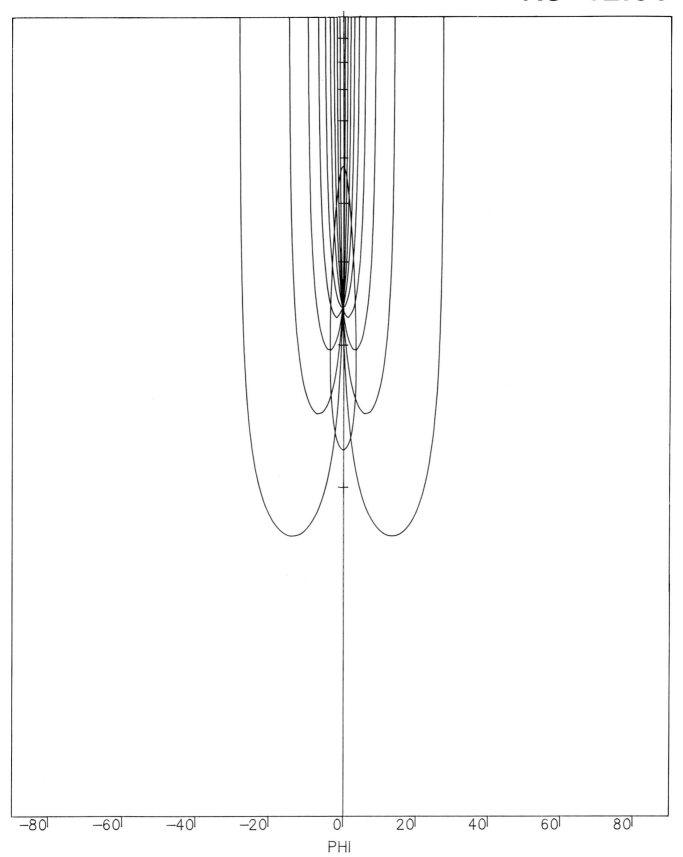

PHI

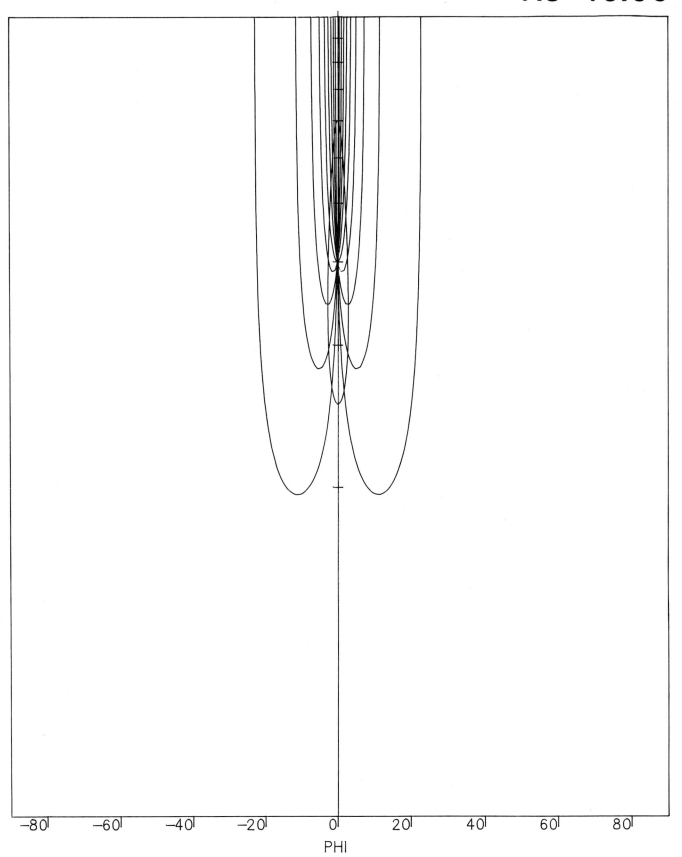

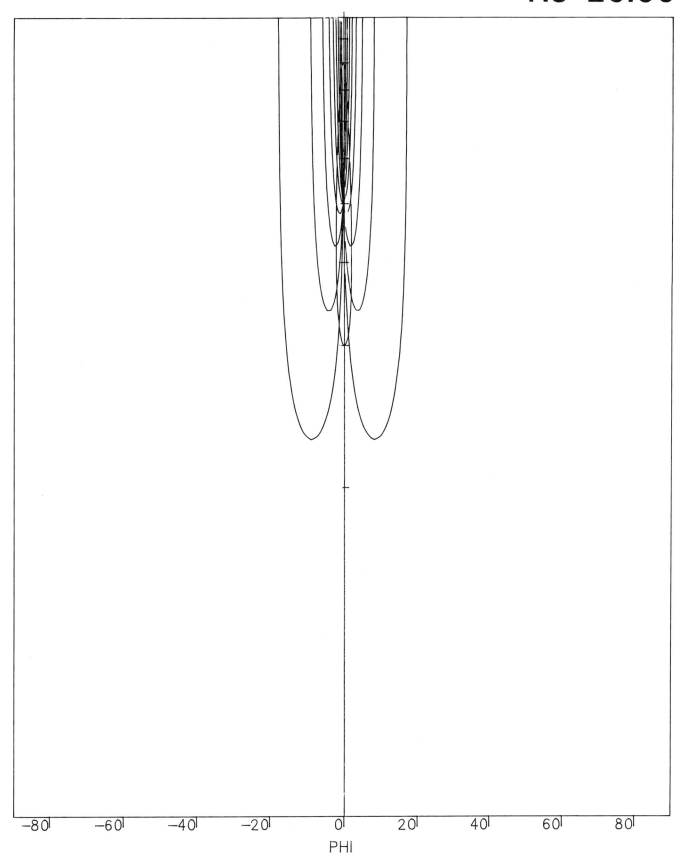

7
Pure Shear Deformation Paths for Various Viscosity Contrasts

The charts presented in this section show the paths followed by markers during their transformation from Ri,θ to Rf,∅. These paths, which are shown as solid lines have a minimum R_f at θ,∅ = 45°. Since the markers rotate towards the principal extension direction each marker moves along its path towards the left (towards lower ∅) during straining. Their progress along the path under a given strain is indicated by the dashed curves which are strain contours drawn for intervals of $0.1\varepsilon_s$. These deformation paths allow R_f and ∅ of a marker to be calculated from Ri,θ and Rs. They are presented here to permit the construction of marker deformation grids of the type shown in Chapter 7 but for different viscosity ratios.

Figure 7.1 shows, as an example, how a constant Ri locus for a deformation grid is constructed. Theta curves are derived in an analogous fashion; by displacing points on a line of constant theta. The θ = 45° curve is given by the form of the strain contours themselves.

Pure shear deformation paths are given for a range of viscosity contrasts. These have been calculated from equations given in section 5.2. Viscosity ratio V = ∞ represents the case of a rigid inclusion in a deformable matrix and V = 0 corresponds to that of an inviscid inclusion (e.g. a vesicle in a lava).

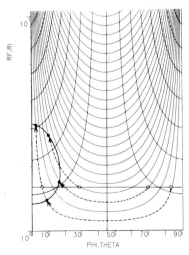

Fig. 7.1 The construction of a constant R_i locus from the pure shear deformation paths

Pure Shear Deformation Paths

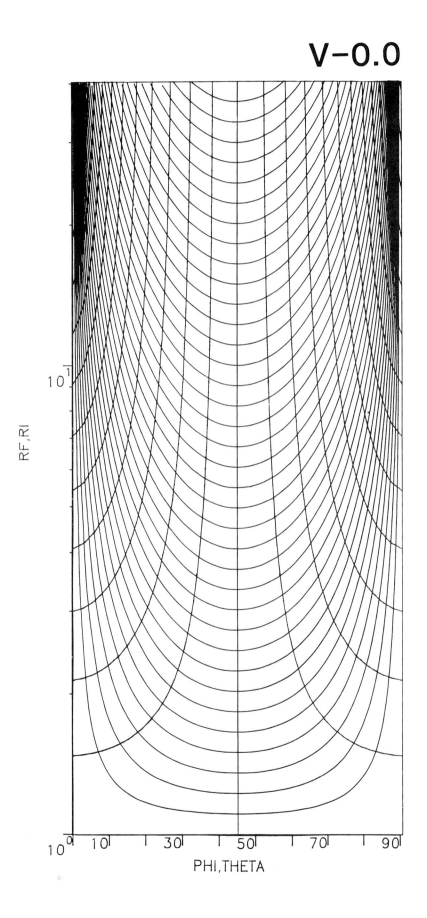

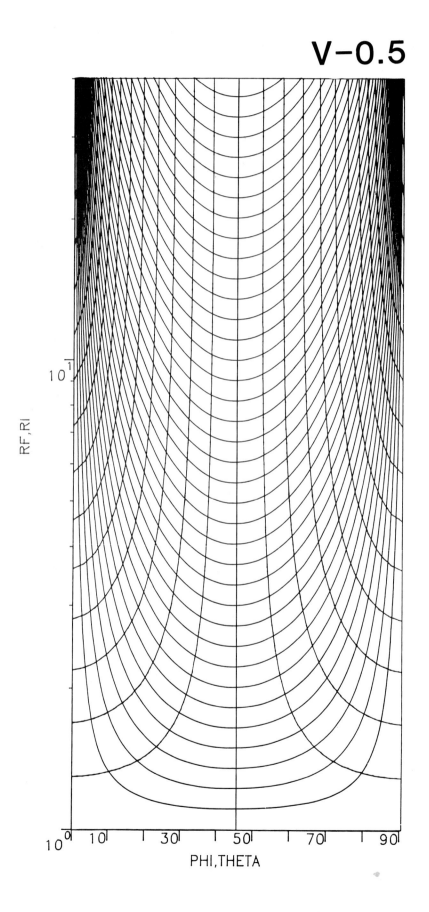

Pure Shear Deformation Paths

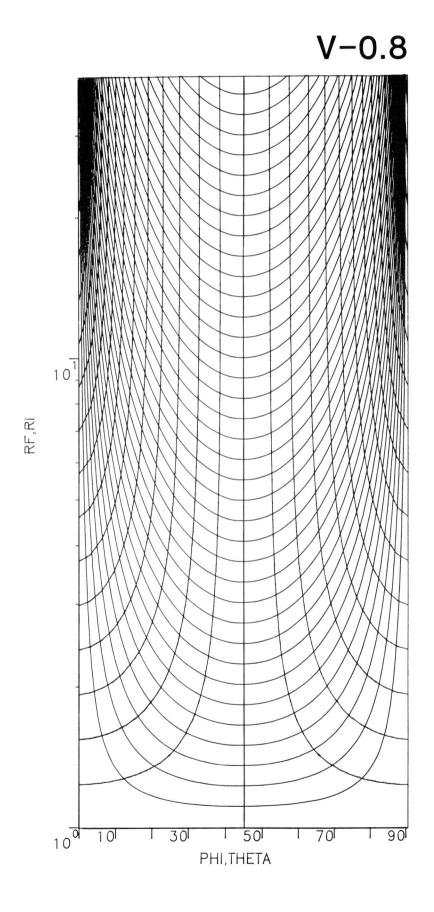

V−0.8

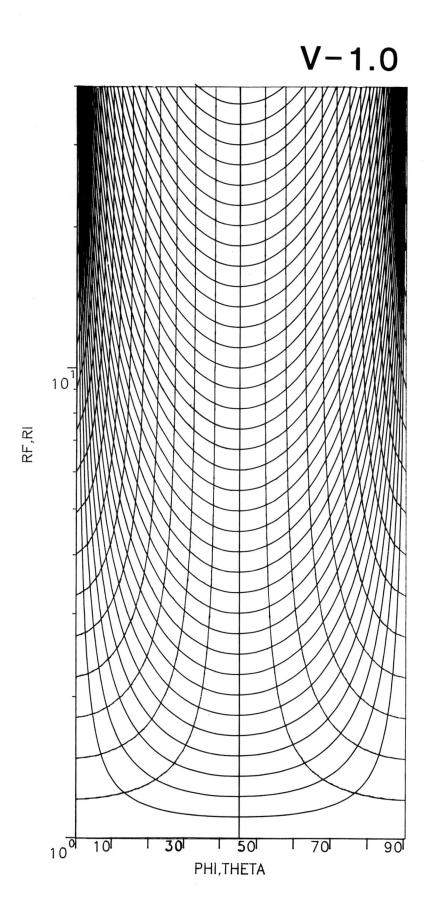

Pure Shear Deformation Paths

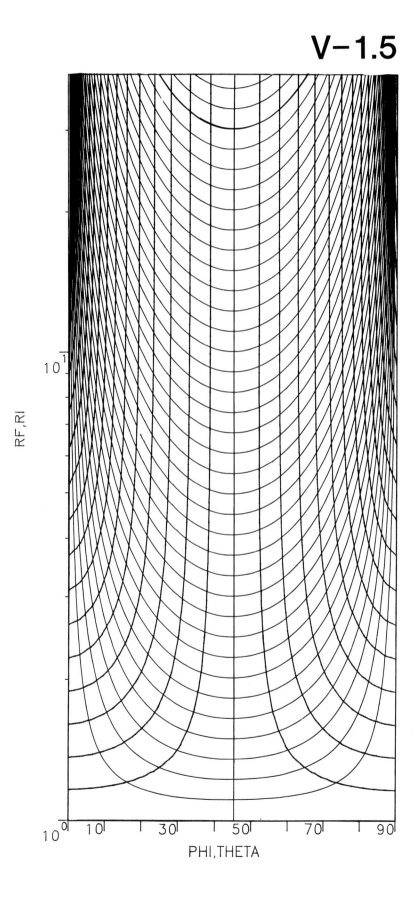

V-1.5

V-2.0

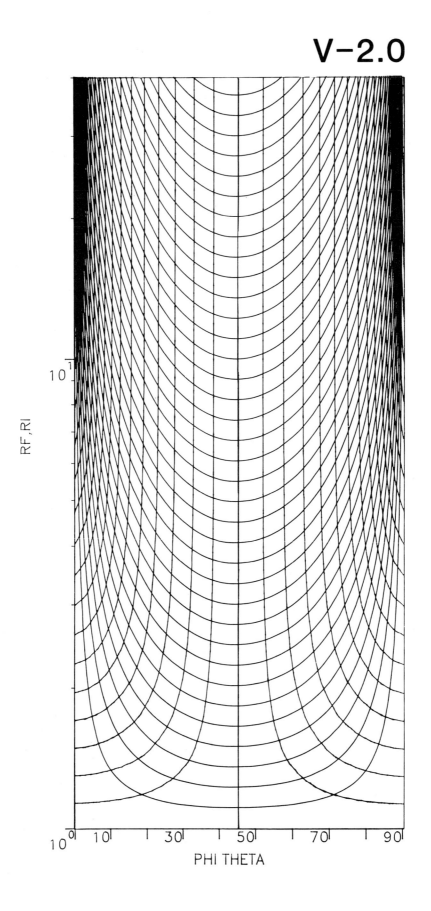

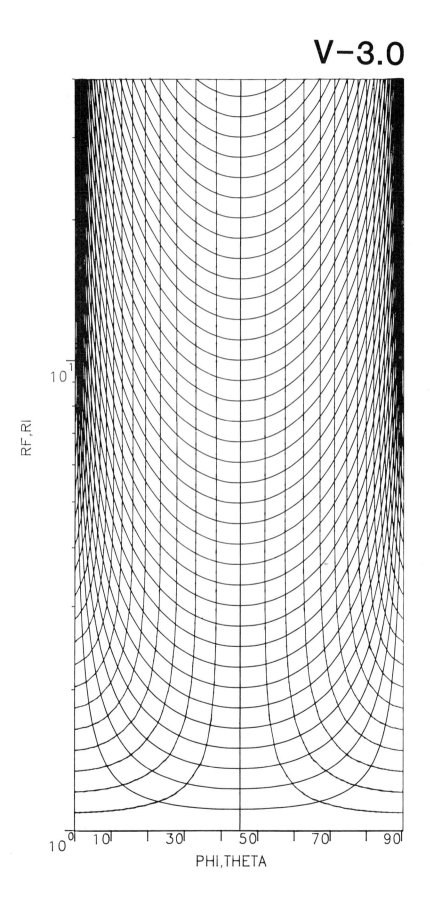

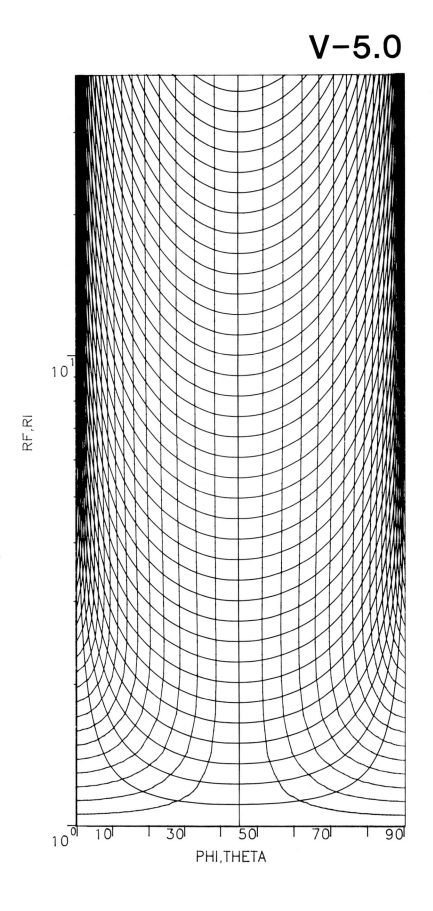

Pure Shear Deformation Paths

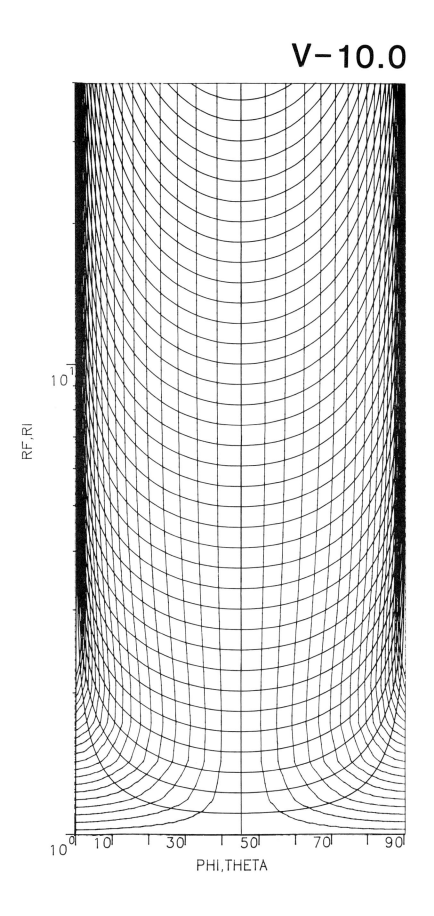

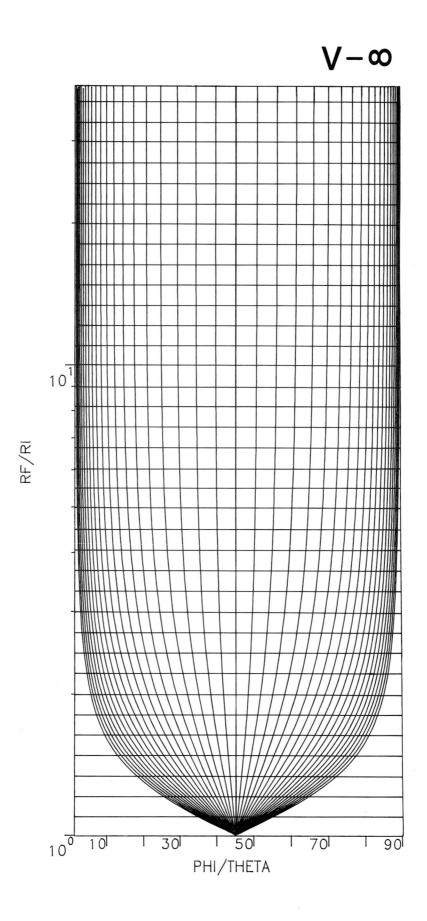

Appendix 1
Basic $R_f/\emptyset$ Equations

The five variables involved in the passive deformation of an ellipse are θ, $\emptyset$, R_i, R_f and R_s. Equations relating these variables are listed below.

$\theta = f(\emptyset, R_i, R_f):$ $\sin 2\theta = \dfrac{R_i(R_f^2-1)}{R_f(R_i^2-1)} \sin 2\emptyset$ (A1.1)

$\theta = f(R_i, R_f, R_s):$ Dunnet (1969) eqn. 29

$\emptyset = f(R_i, R_f, R_s):$ Dunnet (1969) eqns. 28 and 16

$\emptyset = f(\theta, R_i, R_f):$ $\sin 2\emptyset = \dfrac{R_f(R_i^2-1)}{R_i(R_f^2-1)} \sin 2\theta$ (A1.2)

$\emptyset = f(\theta, R_i, R_s):$ Ramsay (1967) eqn. 5.22; Elliott (1970) eqn. 24; Ghosh & Sengupta (1973) eqn. 5; Pfiffner (1980) eqn. 4

$R_i = f(\emptyset, R_f, R_s):$ $\cosh 2\varepsilon_i = \cosh 2\varepsilon_f \cosh 2\varepsilon_s - \cos 2\emptyset \sinh 2\varepsilon_f \sinh 2\varepsilon_s$ (A1.3)
Ribeiro & Possolo (1978) eqn. 9

$R_i = f(\theta, \emptyset, R_f):$ $R_i = \tfrac{1}{2}(R_f - 1/R_f)\sin 2\emptyset \, \text{cosec } 2\theta + \tfrac{1}{2}\left[(R_f - 1R_f)^2 \sin^2 2\emptyset \, \text{cosec}^2 2\theta + 4\right]^{\tfrac{1}{2}}$ (A1.4)

$R_i = f(\theta, \emptyset, R_s):$ $R_i = \left[\dfrac{-2R_s \tan\theta + \tan 2\emptyset \,(1 - R_s^2 \tan^2\theta)}{-2R_s \tan\theta + \tan 2\emptyset \,(R_s^2 - \tan^2\theta)}\right]^{\tfrac{1}{2}}$ (A1.5)

$R_f = f(\theta, \emptyset, R_s):$ Lisle (1977b) p. 385

$R_f = f(\theta, \emptyset, R_i):$ $R_f = \tfrac{1}{2}(R_i - 1/R_i)\sin 2\theta \, \text{cosec } 2\emptyset + \tfrac{1}{2}\left[(R_i - 1/R_i)^2 \sin^2 2\theta \, \text{cosec}^2 2\emptyset + 4\right]^{\tfrac{1}{2}}$ (A1.6)

$R_f = f(\theta, R_i, R_s):$ Elliott (1970) eqn. 23

$R_f = f(\theta, \emptyset, R_i, R_s):$ Ramsay (1967) eqn. 5-27

$R_s = f(\emptyset, R_i, R_f):$ $\sinh 2\varepsilon_s = \dfrac{\sin 2\varepsilon_f \cdot \cos 2\emptyset}{\cosh 2\varepsilon_i} - (\sinh^2 2\varepsilon_i - \sinh^2 2\varepsilon_f \sin^2 2\emptyset)^{\tfrac{1}{2}}$ (A1.7)

$R_s = f(\theta, \emptyset, R_i):$ $\sin 2\varepsilon_s = \tanh 2\varepsilon_i \sin 2\theta \cot 2\emptyset - \sinh 2\varepsilon_i \cos 2\theta$ (A1.8)

Appendix 2

Production of Symmetrical $R_f/\emptyset$ Patterns from Asymmetrical Initial Fabrics

Although the $R_f/\emptyset$ pattern derived from deformation of an initially isotropic fabric of elliptical markers will be symmetric, it is wrong to assume that a symmetric $R_f/\emptyset$ pattern of deformed markers proves that the pre-tectonic fabric was isotropic. Symmetrical $R_f/\emptyset$ fabrics can develop from initial symmetrical fabrics which have been subjected to strains imposed in a symmetrical fashion. Symmetrical $R_f/\emptyset$ fabrics can also be produced by the oblique straining of an asymmetrical initial fabric. The latter possibility can be demonstrated by taking a symmetrical $R_f/\emptyset$ pattern and, with the help of the pure shear deformation paths, destraining it to yield one of the possible starting fabrics (Fig. A2.1). Clearly, although symmetrical fabrics can be theoretically produced in this fashion such fabrics will be rare since they require for their production, a special combination of two variables; the initial fabric and the strain.

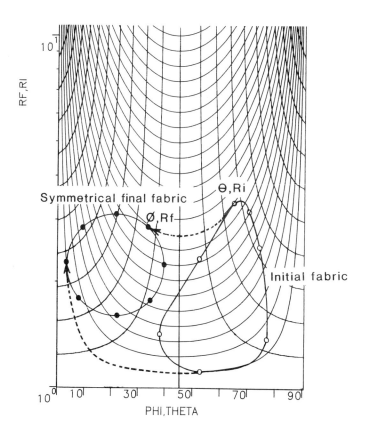

Fig. A2.1

References

Barr, M., and M. P. Coward (1974). A method for the measurement of volume change. Geol. Mag., 111, 293-296.

Bell, A. M. (1981). Strain factorizations from lapilli tuff, English Lake District. Jl. Geol. Soc. London, 138, 463-474.

Bilby, B. A., J. D. Eshelby, and A. K. Kundu (1975). The change of shape of a viscous ellipsoidal region embedded in a slowly deforming matrix having a different viscosity. Tectonophysics, 28, 265-274.

Bilby, B. A., and M. L. Kolbuszewski (1977). The finite deformation of an inhomogeneity in two-dimensional slow viscous incompressible flow. Proc. R. Soc., A355, 335-353.

Bokman, J. (1959). Comparison of two and three dimensional sphericity of sand grains. Bull. geol. Soc. Am., 68, 1689-1692.

Borradaile, G. J. (1979). Strain study of the Caledonides in the Islay region, S.W. Scotland: implications for strain histories and deformation mechanisms in greenschists. Jl. geol. Soc. Lond., 136, 77-88.

Borradaile, G. J. (1984). Strain analysis of passive elliptical markers: success of de-straining methods. J. Struct. Geol., 6, 433-437.

Bouchez, J.-L. (1977). Plastic deformation of quartzites at low temperature in an area of natural strain gradient. Tectonophysics, 39, 25-50.

Boulter, C. A. (1976). Sedimentary fabrics and their relation to strain-analysis methods. Geology, 4, 141-146.

Boulter, C. A. (1983) Compaction-sensitive accretionary lapilli: a means for recognizing soft-sedimentary deformation. Jl. geol. Soc. Lond., 140, 789-794.

Brun, J. P., D. Gapais, and B. Le Theoff (1981). The mantled gneiss domes of Kuopio (Finland): Interfering diapirs. Tectonophysics, 74, 283-304.

Carreras, J., A. Estrada, and S. White (1977). The effects of folding on the c-axis fabrics of a quartz mylonite. Tectonophysics, 39, 3-24.

Chaudhuri, A. K., and A. B. Pal (1980). Pebble strain analysis in the Raghunathpalli Conglomerate of Gangpur Group, Orissa, India. J. geol. Soc. India, 21, 492-496.

Cloos, E. (1947). Oolite deformation in the South Mountain Fold, Maryland. Bull. geol. Soc. Am., 58, 843-913.

Cobbold, P. R. (1980). Compatibility of two-dimensional strains and rotations along strain trajectories. J. Struct. Geol., 2, 379-382.

Cobbold, P. R., and D. Gapais (1983). Pure shear and simple shear of particle/matrix systems with viscosity contrasts. Terra Cognita, 3, (2-3), 247.

Coward, M. P. (1976). Archaean deformation patterns in southern Africa. Phil. Trans. R. Soc. Lond., A283, 313-331.

Cox, S. F. (1981). The stratigraphic and structural setting of the Mt. Lyell Volcanic-hosted sulfide deposits. Econ. Geol., 76, 231-245.

Davidson, D. M. Jr. (1980). Emplacement and deformation of the Archean Saganaga Batholith Vermilion District, N.E. Minnestoa. Tectonophysics, 66, 179-195.

Davidson, D. M. Jr. (1983). Strain analysis of deformed granitic rocks (Helikian), Muskoka District, Ontario. J. Struct. Geol., 5, 181-195.

Debat, P., P. Sirieys, J. Déramond, and J. C. Soula (1975). Paleodeformations d'un massif orthogneissique. Tectonophysics, 28, 159-183.

De Paor, D. G. (1980). Some limitations of the Rf/Ø technique of strain analysis. Tectonophysics, 64, T29-31.

Déramond, J., and D. Litaize (1976). Méthode informatique pour la détermination du taux de déformation. Applications. Bull. Soc. géol. France, série 7, 18, 1423-1433.

Déramond, J., and J.-M. Rambach (1979). Mesure de la déformation dans la nappe de Gavarnie (Pyrénées centrales): interpretation cinématique. Bull Soc. géol. France, 21, 201-211.

Draper, G. (1978). Coaxial pure shear in Jamaica blueschists and deformation associated with subduction. Nature, 275, 735-736.

Dunnet, D. (1969). A technique of finite strain analysis using elliptical particles. Tectonophysics, 7, 117-136.

Dunnet, D., and A. W. B. Siddans (1971). Non-random sedimentary fabrics and their modification by strain. Tectonophysics, 12, 307-325.

Elliott, D. (1970). Determination of finite strain and initial shape from deformed elliptical objects. Bull. geol. Soc. Am., 81, 2221-2236.

Etheridge, M. A., and M. F. Lee (1975). Microstructure of slate from Lady Loretta, Queensland, Australia. Bull. geol. Soc. Amer., 86, 13-22.

Etheridge, M. A., and R. H. Vernon (1981). A deformed polymictic conglomerate; the influence of grain size and composition on the mechanism and rate of deformation. Tectonophysics, 79, 237-254.

Evans, B., M. Rowan, and W. F. Brace (1980). Grain-size sensitive deformation of a stretched conglomerate from Plymouth, Vermont. J. Struct. Geol., 2, 411-424.

Fry, N. (1979). Random point distributions and strain measurement in rocks. Tectonophysics, 60, 89-105.

Gay, N. C. (1968). Pure shear and simple shear deformation of inhomogeneous viscous fluids. 1. Theory. Tectonophysics, 5, 211-234.

Gay, N. C. (1969). The analysis of strain in the Barberton Mountain Land, eastern Transvaal, using deformed pebbles. J. Geol., 77, 377-396.

Geiser, P. A. (1974). Cleavage in some sedimentary rocks of the Central Valley and Ridge Province, Maryland. Bull. geol. Soc. Am., 85, 1399-1412.

Ghosh, S. K., and H. Ramberg (1976). Reorientation of inclusions by combinations of pure shear and simple shear. Tectonophysics, 34, 1-70.

Ghosh, S. K., and S. Sengupta (1973). Compression and simple shear of test models with rigid and deformable inclusions. Tectonophysics, 17, 133-175.

Graham, R. H. (1978). Quantitative deformation studies in the Permian rocks of Alpes-Maritime. Proc. Goguel Symp. (Bur. Rech. Géol. Mines, France), 220-238.

Gray, D. R., and D. W. Durney (1979). Investigations on the mechanical significance of crenulation cleavage. Tectonophysics, 58, 35-79.

Griffiths, J. C. (1967). Scientific Method in the Analysis of Sediments. McGraw-Hill, New York.

Hanna, S. S., and N. Fry (1979). A comparison of methods of strain determination in rocks from southwest Dyfed (Pembrokeshire) and adjacent areas. J. Struct. Geol., 1, 155-162.

Harvey, P. K., and C. C. Ferguson (1981). Directional properties of polygons and their application to finite strain estimation. Tectonophysics, 74, T33-T42.

Helm, D. G., and A. W. B. Siddans (1971). Deformation of a slaty, lapillar tuff in the English Lake District: Discussion. Bull. geol. Soc. Am., 82, 523-531.

Holm, P. E. (1983). The effect of strain heterogeneity on graphical strain analysis methods. Tectonophysics, 95, 101-110.

Holst, T. B. (1982). The role of initial fabric on strain determination from deformed ellipsoidal objects. Tectonophysics, 82, 329-350.

Hutton, D. H. W. (1979). The strain history of a Dalradian slide: using pebbles with low fluctuation in axis orientation. Tectonophysics, 55, 261-273.

Jensen, L. N. (1984). Quartz microfabric of the Laxfordian Canisp Shear Zone, N.W. Scotland. J. Struct. Geol., 6, 293-303.

References

Kelly, T. J., and M. D. Max (1979). A strain section across part of the Caledonian orogen of W. Ireland. In A. L. Harris et al. (Eds.) The Caledonides of British Isles; reviewed. Geol. Soc. Lond., Spec. Publ. 8, 278-280.

Kligfield, R., L. Carmignani, and W. H. Owens (1981). Strain analysis of a Northern Apennine shear zone using deformed marble breccias. J. Struct. Geol., 3, 421-436.

Law, R. D., R. J. Knipe, and H. Dayan (1984). Strain path partitioning with thrust sheets: microstructural and petrofabric evidence from the Moine Thrust Zone at Loch Eriboll, northwest Scotland. J. Struct. Geol., 6, 477-497.

Le Corre, C., and B. Le Theoff (1976). Zonéographie de la déformation finie, de la fabrique et du métamorphisme dans un segment de la chaîne hercynienne armocicaine. Bull. Soc. géol. France, 18, 1435-1442.

Le Theoff, B. (1979). Non-coaxial deformation of elliptical particles. Tectonophysics, 53, T7-T13.

Lisle, R. J. (1977a). Estimation of the tectonic strain ratio from the mean shape of deformed elliptical markers. Geol. Mijnb., 56, 140-144.

Lisle, R. J. (1977b). Clastic grain shape and orientation in relation to cleavage from the Aberystwyth Grits, Wales. Tectonophysics, 39, 381-395.

Lisle, R. J. (1979). Strain analysis using deformed pebbles: The influence of initial pebble shape. Tectonophysics, 60, 263-277.

Lisle, R. J. (1982). Determination of finite strain in deformed planar fabrics. Communic. Serv. Geol. Port., 63, 77-81.

Lisle, R. J., H. E. Rondeel, D. Doorn, J. Brugge, and P. Van de Gaag (1983). Estimation of viscosity contrast and finite strain from deformed elliptical inclusions. J. Struct. Geol., 5, 603-609.

Lisle, R. J., and J. F. Savage (1983). Factors influencing rock competence: data from a Swedish deformed conglomerate. Geol. För. Stockh. Förh, 104, 219-224.

Mancktelow, N. S. (1981). Strain variation between quartz grains of different crystallographic orientation in a naturally deformed metasiltstone. Tectonophysics, 78, 73-84.

Marjoribanks, R. W. (1976). The relation between microfabric and strain in a progressively deformed quartzite sequence from Central Australia. Tectonophysics, 32, 269-293.

Martín Escorza, C., and I. Martín-Montalvo (1980). Determinación puntual del elipsoide de deformación Hercínico que afecto a braquiópodos y nódulos ordovicicos en los Montes de Toledo Occidentales. Estudios geol., 36, 123-129.

Matthews, P. E., R. A. B. Bond, and J. J. Van den Berg (1974). An algebraic method of strain analysis using elliptical markers. Tectonophysics, 24, 31-67.

Miller, D., and G. Oertel (1979). Strain determination from the measurement of pebble shapes: A modification. Tectonophysics, 55, T11-T13.

Milton, N. J. (1980). Determination of the strain ellipsoid from measurements on any three sections. Tectonophysics, 64, T19-T27.

Milton, N. J., and T. J. Chapman (1979). Superposition of plane strain on an initial sedimentary fabric; an example from Laksefjord. J. Struct. Geol., 1, 309-315.

Mitra, S. (1978). Microscopic deformation mechanisms and flow laws in quartzites within the South Mountain anticline. J. Geol., 86, 129-152.

Mukhopadhyay, D. (1973). Strain measurements from deformed quartz grains in the slaty rocks from the Ardennes and the northern Eifel. Tectonophysics, 16, 279-296.

Mukhopadhyay, D., and S. Bhattacharya (1969). A study of pebble deformation in the Precambrian rocks of Singhbhum District, Bihar. J. Geol. Soc. India, 10, 77-87.

Odling, N. E. (1984). Strain analysis of strain path modelling in the Loch Tollie gneisses, Gairloch, N.W. Scotland. J. Struct. Geol., 6, 543-563.

Oertel, G. (1974). Unfolding of an antiform by the reversal of observed strains. Bull. geol. Soc. Am., 85, 445-450.

Oertel, G. (1978). Strain determination from the measurement of pebble shapes. Tectonophysics, 50, 73-78.

Onasch, C. M. (1984). Application of the Rf/Ø technique to elliptical markers deformed by pressure solution. Tectonophysics, 110, 157-165.

Owens, W. H. (1984). The calculation of a best-fit ellipsoid from elliptical sections on arbitrarily orientated planes. J. Struct. Geol., 6, 571-578.

Paterson, S. R. (1983). A comparison of methods used in measuring finite strains from ellipsoidal objects. J. Struct. Geol., 5, 611-618.

Pfiffner, O. A. (1980). Strain analysis in folds (Infrahelvetic Complex, Central Alps). Tectonophysics, 61, 337-362.

Peach, C. J., and R. J. Lisle (1979). A Fortran IV program for the analysis of tectonic strain using deformed elliptical markers. Computers and Geosciences, 5, 325-334.

Percevault, M. N., and P. R. Cobbold (1982). Mathematical removal of regional ductile strains in central Brittany: Evidence of wrench tectonics. Tectonophysics, 82, 317-328.

Ragan, D. M. (1973). Structural Geology, an Introduction to Geometrical Techniques. Wiley, New York.

Ragan, D. M., and M. F. Sheridan (1972). Compaction of the Bishop Tuff, California. Bull. geol. Soc. Am., 83, 95-106.

Ramsay, J. G. (1967). Folding and Fracturing of Rocks. McGraw-Hill, New York.

Ramsay, J. G. (1969). The measurement of strain and displacement in orogenic belts. In P. E. Kent et al. (Eds.), Time and Place in Orogeny. Geol. Soc. Lond., Spec. Publ. 3, 43-81.

Ramsay, J. G., and M. I. Huber (1983). The Techniques of Modern Structural Geology. Academic Press, London.

Ribeiro, A., and A. Possolo (1978). Determination of finite strain in deformed planar fabrics. Communic. Serv. Geol. Port., 63, 77-81.

Ribeiro, A., M. C. Kullberg, and A. Possolo (1983). Finite strain estimation using "anti-clustered" distributions of points. J. Struct. Geol., 5, 233-244.

Roberts, B., and A. W. B. Siddans (1971). Fabric studies in the Llwyd Mawr ignimbrite, Caernarvonshire, North Wales. Tectonophysics, 12, 283-306.

Robin, P.-Y. F. (1977). Determination of geologic strain using randomly oriented strain markers of any shape. Tectonophysics, 42, T7 T16.

Roder, G. H. (1977). Adaption of polygonal strain markers. Tectonophysics, 43, T1-T10.

Roy, S. S., and R. B. Faerseth (1981) Strain analysis of polyphase deformed conglomerate from the Sunnhordland region. Norsk Geol. Tids., 61, 47-58.

Schwerdtner, W. M. (1977). Geometric interpretation of regional strain analyses. Tectonophysics, 39, 515-531.

Schwerdtner, W. M., P. J. Bennet, and W. James (1977). Application of L-S fabric scheme to structural mapping and palaeostrain analysis. Can. J. Earth Sci., 14, 1021-1032.

Sen, R., and A. D. Mukherjee (1972). Strain and shape analysis from the deformed pyrites with a note on their preferred orientations. Geol. Mag., 109, 323-329.

Seymour, D. B., and C. A. Boulter (1979). Tests of computerised strain analysis methods by the analysis of similated deformation of natural unstrained sedimentary fabrics. Tectonophysics, 58, 221-235.

References

Shimamoto, T. (1975). The finite element analysis of the deformation of a viscous spherical body embedded in a viscous medium. J. geol. Soc. Japan, 81, 255-267.

Shimamoto, T., and Y. Ikeda (1976). A simple algebraic method for strain estimation from deformed ellipsoidal objects. Tectonophysics, 36, 315-317.

Siddans, A. W. B. (1976). Deformed rocks and their textures. Phil. Trans. R. Soc. Lond., A283, 43-54.

Siddans, A. W. B. (1979). Deformation, metamorphism and texture development in Permian mudstones of the Glarus Alps. Eclog. geol. Helv., 72, 601-621.

Siddans, A. W. B. (1980). Analysis of three-dimensional homogeneous finite strain using ellipsoidal objects. Tectonophysics, 64, 1-16.

Sparks, R. S. J., and J. V. Wright (1979). Welded air-fall tuffs. Geol. Soc. Amer. Special Paper, 180, 155-166.

Stauffer, M. R., and A. I. Burnett (1979). Down-plunge viewing: a rapid method for estimating the strain ellipsoid for large clasts in deformed rocks. Can. J. Earth Sci., 16, 290-304.

Stephens, M. B. (1975). Pebble strain analysis in the Vojtja conglomerate of central Västerbotten, Sweden. Geol. För. Stockh. Förh., 97, 74-82.

Strömgård, K. E. (1973). Stress distribution during formation of boudinage and pressure shadows. Tectonophysics, 16, 215-248.

Tan, B. K. (1976). Oolite deformation at Windgällen, Canton Uri, Switzerland. Tectonophysics, 31, 157-174.

Tobisch, O. T., R. S. Fiske, S. Sacks, and D. Taniguchi (1977). Strain in metamorphosed volcaniclastic rocks and its bearing on the evolution of orogenic belts. Bull. geol. Soc. Am., 88, 23-40.

Williams, P. F. (1972). Development of metamorphic layering and cleavage in low grade metamorphic rocks at Bermagui, Australia. Am. J. Sci., 272, 1-47.

Windley, B. F., and F. B. Davies (1978). Volcano spacings and lithospheric/crustal thickness in the Archaean. Earth Planet. Sci. Lett., 38, 291-297.

Wheeler, J. (1984). A new plot to display the strain of elliptical markers. J. Struct. Geol., 6, 417-423.

Yu, H., and Y. Zheng (1984). A statistical analysis applied to the Rf/Ø method. Tectonophysics, 110, 151-155.

Author Index

Barr, M.   9, 91
Bell, A.M.   9, 15, 91
Bennet, P.J.   94
Bhattacharya, S.   9, 93
Bilby, B.A.   23, 24, 25, 91
Bokman, J.   16, 91
Bond, R.A.B.   93
Borradaile, G.J.   9, 11, 91
Bouchez, J.-L.   11, 25, 91
Boulter, C.A.   9, 16, 91, 94
Brace, W.F.   2, 92
Brugge, J.   24, 25, 93
Brun, J.P.   9, 91
Burnett, A.I.   9, 16, 95

Carmignani, L.   9, 93
Carreras, J.   25, 91
Chapman, T.J.   8, 93
Chaudhuri, A.K.   9, 91
Cloos, E.   5, 91
Cobbold, P.R.   2, 23, 91, 94
Coward, M.P.   9, 91
Cox, S.F.   91

Davidson, D.M. Jr.   91
Davies, F.B.   1
Dayan, H.   9, 93
Debat, P.   91
De Paor, D.G.   91
Déramond, J.   9, 91, 92
Doorn, D.   24, 25, 93
Draper, G.   92
Dunnet, D.   2, 5, 7, 8, 9, 10, 14, 15, 16, 92
Durney, D.W.   9, 92

Elliott, D.   8, 10, 92
Eshelby, J.D.   23, 91
Estrada, A.   25, 91
Etheridge, M.A.   2, 92
Evans, B.   2, 92

Faerseth, R.B.   9, 94
Ferguson, C.C.   9, 10, 22, 92
Fiske, R.S.   9, 95
Fry, N.   2, 9, 25, 92

Gapais, D.   9, 23, 91
Gay, N.C.   9, 23, 92
Geiser, P.A.   9, 92
Ghosh, S.K.   23, 92
Graham, R.H.   9, 92
Gray, D.R.   9, 92
Griffiths, J.C.   16, 17, 92

Hanna, S.S.   2, 9, 92
Harvey, P.K.   9, 10, 22, 92
Helm, D.G.   16, 92
Holm, P.E.   25, 92
Holst, T.B.   8, 16, 20, 92
Huber, M.I.   9, 14, 94
Hutton, D.H.W.   9, 92

Ikeda, Y.   11, 95

James, W.   94
Jenson, L.N.   9, 92

Kelly, T.J.   9, 93
Kligfield, R.   9, 93
Knipe, R.J.   9, 93
Kolbuszewski, M.L.   24, 25, 91
Kullberg, M.C.   94
Kundu, A.K.   23, 91

Law, R.D.   9, 93
Le Corre, C.   9, 93
Lee, M.F.   92
Le Theoff, B.   9, 91, 93
Lisle, R.J.   2, 6, 11, 15, 16, 24, 25, 93, 94
Litaize, D.   91

Mancktelow, N.S.   9, 25, 93
Marjoribanks, R.W.   9, 25, 93
Martín Escorza, C.   9, 16, 93
Martín-Montalvo   9, 16, 93
Matthews, P.E.   93
Max, M.D.   9, 93
Miller, D.   11, 93
Milton, N.J.   8, 11, 93
Mitra, S.   1, 93
Mukherjee, A.D.   9, 94
Mukhopadhyay, D.   8, 9, 94

Odling, N.E.   9, 94
Oertel, G.   2, 11, 93, 94
Onasch, C.M.   25, 94
Owens, W.H.   8, 9, 11, 93, 94

Pal, A.B.   9, 91
Paterson, S.R.   2, 8, 10, 94
Peach, C.J.   94
Percevault, M.N.   2, 94
Pfiffner, O.A.   9, 16, 94
Possolo, A.   9, 94

Ragan, D.M.   2, 94
Rambach, J.-M.   9, 92
Ramberg, H.   23, 92
Ramsay, J.G.   1, 2, 3, 5, 9, 10, 11, 14, 23, 94
Ribeiro, A.   9, 94
Roberts, B.   9, 94
Robin, P.-Y.F.   9, 16, 94
Roder, G.H.   10, 94
Rondeel, H.E.   24, 25, 93
Rowan, M.   2, 92
Roy, S.S.   9, 94

Sacks, S.   9, 95
Savage, J.F.   2, 10, 93
Schwerdtner, W.M.   2, 94
Sen, R.   9, 94
Sengupta, S.   23, 92

Seymour, D.B.   16, 94
Sheridan, M.F.   2
Shimamoto, T.   11, 23, 95
Siddans, A.W.B.   2, 7, 9, 10, 11, 14, 15, 16, 92, 94, 95
Sirieys, J.   91
Soula, J.C.   91
Sparks, R.S.J.   2, 9, 16, 95
Stauffer, M.R.   9, 16, 95
Stephens, M.B.   9, 95
Strömgård, K.E.   23, 95

Tan, B.K.   8, 9, 95
Taniguchi, D.   9, 95
Tobisch, O.T.   9, 95

Van de Gaag, P.   24, 25, 93
Van den Berg, J.J.   93
Vernon, R.H.   2, 92

Wheeler, J.   9, 16, 95
White, S.   25, 91
Williams, P.F.   25, 95
Windley, B.F.   1, 95
Wright, J.V.   2, 9, 16, 95

Yu, H.   95

Zheng, Y.   95

Subject Index

Accretionary lapilli 9, 16
Amygdales 9
Axial ratios of deformed markers 4
  extreme ratios 5, 16

Bedding-parallel fabric 20
Biotite 9
Boudinage 9, 20
Breccia 9
Burrows 9

Cataclasis 8
Chi-square test 16
Cleavage 22
Compaction 2
Competence contrasts 13, 23
Concretions 20
Conglomerate 9, 16
Cordierite 9, 16
Crystal-plastic deformation 10
Crystallographic fabrics 2
Curve-fitting criteria 22

Data, amount of 10
  plotting 11
Data collection 8
Deformation, homogeneous 3
  pure shear 24
  history 25
Desiccation cracks 9

Elliptical markers, deformation of 3, 4
  constant eccentricity 4
  extreme shapes 5
  constant orientation 6

Fabrics, crystallographic 2
  initial 15, 16
  sedimentary 16, 20
  symmetrical 20
Feldspar clasts 9
Flow chart for strain
  determination 12, 13
Fluctuation 5, 6, 16, 25
  curves of 6
Foliation 2, 23
Fossils 10

Grain growth 8
Grain size 2
Grit 16
Harmonic mean 14, 15, 18
Heterogeneous strain

Isym test 14
  critical values 15

Marker deformation grids 7, 25, 26
Marker separation method 25
Mean, vector 14
  harmonic 14, 18
Measurements 8
Muscovite 9

Nodules in slates 9

Oncalites 9
Oolites 9, 16, 25

Pebbles 16
Pinch-and-swell structure 24
Pipes 9, 16
Preferres orientations 12
  unimodal 12
  uniform 15
  pre-tectonic 19, 20
  symmetrical 20
Pressure shadows 23
Pressure solution 11, 25
Pumice 9, 16
Pure shear 24
  pure shear deformation paths 25

Quartz, grains 9, 16
  aggregates 9

Reduction spots 9
$R_f/\emptyset$ analyses 15, 16
Rheology 2
Rutile, as strain marker 1

Sandstone 16
Sedimentary structures 1
Shape factor 2
Slate fragments 16
Slaty cleavage 22
Standard curves 12, 15, 26
  fitting data to 16, 17
Strain, three dimensional 8, 11
  determination 12-22
  heterogeneous 23, 25
Strain ellipse 3, 22
Strain ellipsoid 2, 22
Strain markers 1
  ooids 2
  subelliptical 8
  parallelogrammatical 8, 10
  rhombic 10
  polygonal 10
  anisotropic 25
Stratigraphic thicknesses, restoration 1
Symmetry of $R_f/\emptyset$ diagrams 14
Symmetry test for $R_f/\emptyset$ diagrams 15

Theta curves 6
Tuff 16

Varioles 9
Vector mean 14
Viscosity contrasts 12, 23-25

Xenoliths 9